Decolonizing Peace

Decolonizing Peace

By Victoria C. Fontan

Published by World Dignity University Press, an imprint of Dignity Press
16 Northview Court
Lake Oswego, OR 97035, USA
www.dignitypress.org
Book website: www.dignitypress.org/decolonizing-peace

Book design by Uli Spalthoff
Cover based on the painting "Vientos Alicios", from Isidro Con Wong, 2012
Printed on paper from environmentally managed forestry:
www.lightningsource.com/chainofcustody

ISBN 978-1-937570-15-6
Also available as EPUB: ISBN 978-1-937570-16-3
and Kindle eBook: ISBN 978-1-937570-17-0

For Elias Cheboud and
Mahmoud el-Zain Hamid

For Hermine and Jean-Philippe Lafont

Contents

Acknowledgements 11

Introduction 13

I. The Case for Decolonizing Peace 23

Huda and Sajeeda 25
Human Trafficking and International Law 26
Abu Baker, Buk and Koran 28
The Liberal Peace Paradigm 31
The Field beyond Right and Wrong 34
No solutions 36
Liberal Peace and its Enabling Great Caveat 38
The Peace Universalism 39
Decolonizing Peace 42
A Different Lens 43
Decolonizing Methodologies: Teaching and Writing to Transgress 45
Conclusion 46

II. A Leap of Faith 49

"If God Existed" 50
Neither Mara, nor Slave 53
On Revolutions 55
On the Immanence of Peace 59
What Happens in the "Field" and Whose Peace Are We Working for? 63
Conclusion 68

III. Anomalies and the Invisible 71

The Lebanese Hezbollah, the Anomaly's Visible 72
Complex Adaptive Systems 75
The Formation and Nurturing of a Strange Attractor 77
Hezbollah's Invisible: The wilāyat al-faqīh 78
Hezbollah's Transformation As a Complex Adaptive System 82
The Anomaly's Sustainability As a Complex Adaptive System: A Given? 85
Letters from Abbottabad 86
An Ordinary Woman, Almost 91
The Great NGO Scheme 93
Gulabi 95
An Open Circle of Help 96
A Self-Organized Chaos 98
Formative Events and the Panarchy of Living Systems 101
Revisiting Sustainable Peace 106
Including the Anomalies into the Formation of Peace 109
Complex Adaptive Anomalies 111

IV. A Journey Through the "Sacred" 113

Addiction 115
Double Bind and Cybernetics 118
Cybernetics of Peace 122
From Tahrir to Wall Street 124
Dr. Jekyll and Mr. Hyde 127
Bateson's "Sacred" Curse? 130
Kinesthetics of Peace 132
People Are the Best Resources We Have 134
Connecting Two Worlds In the Same Town 137
Conclusion 141

V. Emergence 143

From Networks to Communities of Practice 145
Yin and Yang 148
The Do-ocracy of the Hive 152
Obsolescence 156
Systems of Influence: When the Anomaly Becomes the Epistemology 158
Dialogue as an Amplifier 162
Conclusion 165

Epilogue 167

Bibliography 175

Notes 187

Index 214

Acknowledgements

I acknowledge the Government of the Netherlands for the priceless exposure to the world that its project funding has granted me over the past five years, as well as Thomas Klompmaker for managing it so efficiently. I thank the University for Peace for giving me an academic home and granting me full academic freedom over the years.

I thank Anne Robert and the Universidad de La Salle doctoral team; my colleagues: Dery Dyer, Jacqueline Gillet, Jan Hurwitch, Steve Kogel, and Jim Molloy; Santiago Slabodsky for his encouragements throughout; Damien Larramendy for being there at the right time; David Walker at HMT Insurance for granting me peace of mind anywhere in the world; Irene Paniagua for her illustrations, and finally Ross Ryan for his editing. Many colleagues have reviewed some chapters of my initial manuscript, and I am grateful to them for their invaluable feedback: Joseph Alagha, Yael Efron, Cynthia Enloe, Laurel Gaylor, and Oliver Richmond. All remaining errors are mine alone.

Dignity Press has allowed me to fully be myself while writing this book, for not having to fit into a contrived academic box. This book would never have taken its present form if it weren't for Linda Hartling, Evelin Lindner, and Uli Spalthoff.

Finally, I thank all the people who shared their lives with me and agreed to let me write their stories. Special thanks go to Geovanni Morales, who has inspired me deeply.

"There is no utopia to strive for
–there is only now"

Peter Fein

Introduction

There has been no single formative event that led to the writing of this book. Rather, it is the random collection of many experiences, in the field and in the classroom, that has led me to challenge the foundations of peace and conflict studies as I know them. Over the years, the processing of past events has allowed me to assemble a puzzle that is different to the pre-established one I have been expected to teach my students at former times. My peace and conflict puzzle is constantly in the making, it has no original or final shape, and it bears too many dimensions to be seen as the sum of its original pieces.

In this book, I will question the pre-established assumptions that exist in peace and conflict studies, and bring forward an alternative epistemology that relates to unconventional initiatives for peace that exist in various parts of the world. I will seek to deconstruct "peace" as we claim to perform it, and ask some crucial questions, not only about the efficiency of our work, but about the structure of our system itself, in an attempt to identify how and why it keeps failing the populations we often claim to be protecting, helping, or assisting. Who are we protecting when we foster the resurgence of human trafficking in post-conflict situations? Who are we helping when we send our most inexperienced practitioners to the toughest regions of the world? Who are we assisting when just a fraction of the money we send actually reaches the ground?

The first illustration that I always use to raise students' awareness of the current challenges faced by peace and conflict studies is my experience in post-conflict Bosnia-Herzegovina in 2001. I had just turned twenty-five, and I was appointed as Democratization Officer and Acting Head of Field Office by the Organization for Security and Co-operation in Europe (OSCE). At the time, I was a PhD candidate at the University of Limerick, Ireland, and needed funds to carry on with my studies. I applied online to the OSCE's democratization team, and was granted a telephone interview soon after. Two factors seemed important to my recruitment: my experience in interacting with civil society organizations, and my willingness to be deployed at short notice in challenging environments. As I ached to gain experience in what I used to refer to as the "field", I jumped at the opportunity to be deployed in an environment that, I thought, would finally make a peacebuilder out of me, and which would give me some of the field credentials necessary to be considered a successful peace and conflict studies scholar. As I arrived in Sarajevo, I learned that I would be deployed in Drvar, North-Western Bosnia, a small town full of problems that our office, as part of handful of international organizations present in the area, would be expected to solve. It was made clear to me that Drvar was no one's first choice as an assignment; in fact, it seemed that no one wanted to be deployed there. Drvar was considered too difficult an environment to work in, as most of its population was hostile to any international presence, and had been resisting peacebuilding initiatives on various occasions since the Dayton Peace Agreement of 1995. Tensions remained high between different parties, and economic/political developments were inexistent. Last, but very importantly for some, I was told that Drvar was also remote from the sea, Sarajevo, or any "civilized" environment. Since it was expected that hard-working peacebuilders should be able go away for the weekends to a well deserved "rest and relaxation," colleagues in Sarajevo joked with me that I was to be deployed to "the armpit of Bosnia".

I did not mind any of this, as I considered this mission to be the launching pad for my career. There was, however, one unexpected drawback to my deployment that caused me many sleepless nights: I did not know what the OSCE meant by "democratization". What was expected of me as Democratization Officer? Since most dictatorships in the world refer to themselves as democracies, I was aware that the term was as elusive as its many avatars. When I raised the obvious question to my recruiter in Sarajevo, I was told that I should go on the mission's website to study my mandate and pillars, and I was assured that my experienced twenty-eight year-old democratization colleague in the neighboring town of Livno would coach me as I arrived to my area of responsibility. Being "experienced" in that particular context meant having a German passport and two years of previous experience as a democratization officer. As my experienced colleague was on rest and relaxation from her most difficult life as a peacebuilder, her assistant, a 50 year-old "local" woman named Ankica, welcomed me to town. She had spent all her life in Livno, and knew the system inside and out. She was respected by everyone in town, and walked me through what my duties were to be. As I gasped at the immensity of my mandate, as well as the scarce means I would be given to honor it, I realized that I was expected to coach politicians in their 50s and 60s on democratization, inspire youth actors who would never be afforded a fraction of the many opportunities that life had given me, and "empower" women whose daily lives I knew or understood nothing about. I was expected to coach all those people on "democratization" just because I happened to carry a French passport. As I was still convinced that my country was a beacon of human rights and democracy, I was ready to take on the challenge. Still, I realized that there was an exchange rate, a categorical difference, between the international staff, paid for the hardships of living in the "bush" far away from "civilization", and the locals, the backbone of our daily activities, who were paid a fraction of our salaries. Just because I had been born on the right side of the fence, I was expected to come up with solutions for the lives of communities that had experienced incredibly traumatic, earth-shattering violence. If I was supposed to represent any sort of hope, the "international community" was playing a cruel joke on them.

Since my only experience was in abstract academia, I set to map-out the situation that I had inherited as soon as I arrived in Drvar. In a paper entitled "Multi-Track diplomacy in Bosnia-Herzegovina: Post-Conflict rehabilitation in Drvar," I applied my academic training to identifying vulnerable/at-risk groups in my area, as well as ensuing priorities for sustainable post-conflict rehabilitation.[1] When I matched this analysis to my mandate, I realized that the budget that I had to address these priorities was barely enough to decorate workshop tables with pretty plastic plants and flowers. Where was the disconnect? Why was Drvar considered to be a "post-conflict environment", when all that I could see on the ground was another conflict in the making?

Drvar may have been considered to be "the armpit of Bosnia" by some, but it also turned out to be the most complex political situation in the entire mission: a true peace and conflict studies microcosm. Originally a Serb town, it had been ethnically cleansed of most of its Serbian population in September 1995. In anticipation of Bosnia's General Framework Agreement for Peace, more commonly knows as the Dayton Agreement, which was to be negotiated by all parties to the conflict in December 1995, Croatian leadership had rushed to alter the ethnic make-up of this strategic town. Since it was established that Bosnia-Herzegovina would be partitioned, and that Drvar would fall within the remit of the Bosnian-Croat federation, harboring an ethnic majority of Muslims Bosniacs and Croats, the ethnic cleansing of Drvar was a last minute, yet well-orchestrated effort. As the local Croatian leadership, spearheaded by Drago Tokmadjia, sought to gain as much territory as possible, all strategic businesses and public institutions in town were annexed by its Croatian elite.[2] Houses were emptied of their Serb inhabitants, and repopulated by Croats who themselves had been internally displaced from other parts of Bosnia-Herzegovina.

As a Croat, one would have access to all private and public services such as telecommunications, health-care, schooling, military pension, security, etc.; while as a Serb, one would find it extremely difficult, if not impossible, to re-settle in town, due to a complete lack of services and community support.[3] Should a Serb family decide to return, no security would be provided for any of its members, as the local police and army presence were all Croat. Worse, acts of intimidation on the part of local police were frequently reported to the "international community". Under these inhospitable premises, the main mission of our office, more specifically its human rights department, was to accompany the implementation of property law, so that Serbs would be facilitated in the return to their homes.[4] This seemed much more realistic than the inapplicable democratization mandate that Sarajevo had dictated to me without ever, it seemed, having stepped a foot in Drvar. While the property claim process had been initiated five years earlier all throughout

Bosnia-Herzegovina, we soon came to realize that it had come to a halt in Drvar due to a combination of local hostility toward this particular part of our mandate and excessive caution on the part of our superiors. Six years into the implementation of the Dayton Agreement, evictions were at a standstill in town, with very little international commitment to honor this part of the OSCE mandate. Why?

In 1998, the newly elected Serb mayor of Drvar, Mile Marceta, was thrown out of his office window by an angry mob as part of a violent demonstration organized by the local Croat elite to pressure the international community into invalidating the latest municipal election results. As Marceta was left for dead on the street pavement, he was recuperated by Canadian NATO forces *in-extremis*.[5] Marceta had been elected as a result of the organization and observation of free and fair elections, as part of a healthy post-conflict democratization process. Surely, the tragic events that almost killed him would be severely sanctioned by the International Community? The plenipotentiary international ruler of the area at the time was Peter Chappel, a former humanitarian aid convoy truck driver. Chappel was the local Special Envoy of the Office of the High Representative, an *ad hoc* instrument created by the international community to "monitor the implementation of the peace settlement."[6] In effect, the OHR was assuming the authority of the Bosnia-Herzegovina state, and had all powers to adequately ensure that such a situation would never be allowed to happen again. Chappel's decision was drastic. He immediately removed Marceta from office, arguing that it was his election that had triggered unrest in Drvar. He also sanctioned the alleged organizer of the demonstrations, Drago Tokmadjia, with ineligibility. After this decision, the municipality went back to the Croat community, whose *de facto* leader remained Tokmadjia until we arrived in town.

Our immediate management considered that events such as the 1998 riots ought to be avoided at all costs in the future. Since threats of a similar breakout were made as soon as we took office, we realized that the local elite were using blatant intimidation in an effort to stop refugee returns. Should we have given in to these threats, none of our mandate could ever have been implemented. We would have become ornamental peacebuilders in our own offices. At the same time, my newly appointed human rights colleague, an Italian lawyer, was receiving Serb home owners on a daily basis, who plead to be allowed to return to their homes before the harsh winter broke out. Some of them lived in caravans, hounded by wolves in the countryside. Others had makeshift lodgings in covered parking spots. Overall, their living conditions were gruesome, while their own homes were occupied by others. Some were so old that we knew they might die before ever returning to their homes. Others had managed to repossess their homes, only to find them burned or booby-trapped. We felt accountable to those individuals. We wanted to bring them the good news that after six years of standstill, we would finally facilitate their return.

So long as such a return could be avoided, Mr Tokmadjia and other businessmen in town were free to maintain their control over the area, most noticeably through the illegal logging activities of the Croat-owned company Finvest, which was no doubt benefitting them personally. When we arrived in town, my human rights counterpart and I were warned by our Regional Centre Director, Ms. C., that we would not be able to achieve much. She encouraged us to be patient, and promised that she would give us the go ahead on our tasks soon. Interestingly enough, both my human rights colleague and I were deployed on our first assignment, with no "field" previous experience. Was this intentional? Was the toughest spot in Bosnia-Herzegovina to be manned with first-timers? Ms. C. stressed that she had handpicked us for this difficult task, that she had high hopes for us in the organization. While we were certainly flattered by her trust in our abilities, we also felt accountable to our local interlocutors, waiting to repossess their homes as soon as possible. Our commitment to them superseded our desire to please our immediate hierarchy. After writing numerous reports to Sarajevo, bypassing Ms. C.'s authority, we finally obtained the green light to embark on honoring the office's property law mandate. As we initiated our office's most important task, the anniversary of the ethnic cleansing of the town was celebrated by arson, which resulted in an immense forest fire that was only contained by heavy rain after days of burning. The warning was chilling: if the Croat community were made to leave the area, there would be no town standing to be repossessed.

Ms. C. informed us that since we were to initiate the refugee return process, she would personally come to talk Tokmadjia into co-operating with us, or at least not making our job more difficult than it had to be. A meeting was set for late September. As Regional Centre Director, she was to make him understand that the eviction process of hundreds of Croatian families could not be delayed any longer. Ms. C. claimed that she was an experienced negotiator. Her high position in the organization certainly vouched for this. As the time of the meeting approached, she explained that she used to sell used airplanes to African dictators. According to her perception, this would make her the toughest and most experienced negotiator in town. She was very friendly while we waited to meet Tokmadjia at the terrace of the largest restaurant in town – there were only two – and she encouraged us to be upbeat, and to "watch and learn." After his arrival and informal introductions, the atmosphere of the meeting rapidly became sexually tense. Ms. C. and Tokmadjia openly began to exchange inappropriate jokes and comments. My human rights colleague and I were feeling increasingly uncomfortable. We assumed that she knew what she was doing, since she was our boss. We were all pressured to start drinking whisky together. After we left to attend pressing duties, Ms. C. ended this particular meeting on Tokmadjia's lap, hand-feeding him pepperoni, in view of the entire town. Our translator then called us to let us know that both had initiated another meeting, in the private confines of a room of the Bastasi hotel, in the village next to Drvar. The next day, Tokmadjia bragged around town to have kept Ms. C.'s underwear as a trophy.

Our Serbian staff came to the office with many questions that we could not answer. How were we to implement any mandate, if our boss had openly sided with one party to the conflict? What was to become of our efforts, our commitment to those returnees without homes? What was the political impact of such a strategy? All morning, my human rights colleague and I paced around the office, asking ourselves what Ms. C.'s strategy could be, if there was one at all. We had watched something very shocking, and had no intention of learning any of it! We could not grasp what had just occurred, and we questioned our staff to know exactly what had happened after we left the restaurant. The advantage of being first timers in a job like ours was that we were unaware of the politics of our organization. We were also idealists, strong believers of the ideals that had brought us to our respective positions. We thus decided to blow the whistle, one month into our job. It was a hard decision to make, as we both knew that our careers might not survive it.

An OSCE internal investigator, a former policeman from the US, was sent by our Sarajevo headquarters to interrogate us. As he arrived, he let us know very clearly that Ms. C. was a "colorful person," very well connected to the Berlusconi administration, implying that we ought to put what had happened behind us. It was clearly insinuated that the investigation was going to turn against us if we spoke out. Repeated phone calls and visits were made to intimidate us, and have us dampen our testimonies. Still, our young aged idealism prevailed. While our organization was not supporting us, NATO, whose position in town had also been weakened by the Ms. C.'s behavior, decided to step in. Their leadership in Sarajevo contacted our organizational leadership, and after much struggle, Ms. C. was asked to resign. By then, I had already left the mission, having been awarded a PhD Fellowship that prevented me from engaging in any paid employment, and more importantly, having realized that I never wanted to work for such an organization again. Some were made to pay for Ms. C.'s resignation, including my immediate supervisor, Senior Democratization Officer Michaela Bergman, who had supported me throughout. Miraculously, my human rights colleague was confirmed in his position, after refusing to be transferred to another field office, with the full support of his Human Rights hierarchy. It was later alleged that Ms. C. had business links with Tokmadjia, alongside OHR Representative Mr. C., but there was no official enquiry to verify this as she had already left the mission.

Upon Ms. C.'s departure, some colleagues wrote to me, expressing their support and also making further allegations as to her pattern of wrongdoings and shady political deals. In a post preceding her appointment in Mostar, as Regional Centre Director in Tuzla, allegedly she had been given property by a local politician in exchange for political favors.[7] To my dismay, three years later, she reappeared on the post-conflict scene, this time as US-led Coalition Provisional Authority Governorate

Co-ordinator of the Dhi Qar province, in *Iraq*.[8] A very close ally to Italian President Silvio Berlusconi, she had *de facto* been given a promotion.[9] After this stint, she came under media scrutiny for vote rigging as part of her candidacy for the Italian Senate.[10] Some years later, she was appointed special envoy of the Italian Ministry of Foreign Affairs to Darfur. Under those auspices, she came to the forefront of a scandal over the non-appropriation of funds raised in Italy for a hospital in the town of Nyala.[11]

As I witnessed Ms. C.'s ascension of the diplomatic ladder, I have been left to watch as one individual damaged a peace mission, ruining the expectations of local individuals, only to be transferred to another environment where other individuals were equally short-changed. What would I find if I ran detailed interviews on her tenure in both Iraq and Darfur? This case reminds me of how the Catholic Church would re-settle known pedophiles to other parishes, as it continuously refused to act on its own demons. Is the trade of peace, and the discipline of peace and conflict studies, guilty of the same charges? Are good intentions enough to make peace? Is peace, the way it is being appropriated by the system, part of the wider conflict problem? Is Ms. C. an exception to the rule or the product of a problematic system? In short, can we possibly do better?

In this book, I will seek to address all the questions that this first encounter with the peace industry, and subsequent peace and conflict research, has raised not only in me but also in my colleagues throughout the world. This book will be divided into several parts, first deconstructing the discipline and practice of peace and conflict studies, while making a case for a decolonizing approach to peace; then looking at the implications of an already existing paradigm shift; followed by an analysis of the different anomalies that can emerge as part of a dysfunctional political or peace system. An analysis of alternative understandings of the "field" realities from a theoretical perspective will follow, as well as a snapshot ethnography of various realities I have encountered throughout the world. In the last two chapters, alternative pathways for decolonizing peace will be introduced with an additional theoretical understanding, to be followed by a systemic wrapping up of the individual and collective modalities of decolonizing peace.

This book does not aspire to be a conventional academic book, both from the perspective of the narrative and style that it will keep utilizing, as well as the concepts that will be introduced to process the real life examples that are developed throughout. It will use a trans-disciplinary epistemology to make sense of a complex "field" reality. The idea of this book is not only to establish an alternative paradigm from which to understand peace and conflict studies, but also to use ongoing initiatives and examples to decipher a parallel reality both epistemologically and theoretically. The purpose of this book is not to dwell on the tools of conventional, liberal peace, but on the invisible aspects of the territory of peace that are left behind by the traditional tools and epistemology of peace. Lastly, this book will not come into opposition with conventional concepts or theories. Rather, it seeks to complement and enrich them.

In order to assess the invisible aspects of the peace territory, let us first look at the most visible ones, and the many ramifications that an incomplete understanding of peace, as illustrated in my Bosnia example, may have on the ground.

I. The Case for Decolonizing Peace

In the past few years, rape in Eastern Congo has been the object of intense media coverage. This has contributed to the assistance and treatment of thousands of women and girls, taken from the front lines into safe houses in different urban centers of the region.[12] While the exact number of victims is difficult to gather, UN officials have referred to the Democratic Republic of Congo (DR Congo) as the "worst place in the world in terms of sexual violence."[13] There is one aspect of sexual violence, however, that the UN mission in DR Congo would rather abstain from mentioning: the rape of children and internally trafficked young women by some of its mission members within the towns of Bukavu, Goma and Uvira.[14] According to Congolese law, sexual intercourse involving children amounts to rape.[15] Yet, peacekeepers regularly pay for sex with children and women in and around some nightclubs and hotels in Kivu Province of DR Congo. While the aforementioned UN narrative paints sexual violence in DR Congo as a problem to be solved by the intervention of a benevolent savior, one may well consider the presence of this "savior" as a contributor of the current insecurity in DR Congo.

Amidst the controversy about Greg Mortensen's *Three Cups of Tea*, it is important to question, not only the motives and good intentions of self-proclaimed peacebuilders, but the systemic structures that allow for human rights abuses, embezzlement, and corruption to take place as part of peace missions.[16] Are those abuses the product of "a few bad apples", an exception to the rule, as each probe reluctantly carried out tends to tell us, or are the epistemological foundations of peace missions themselves responsible? Has our understanding of peace become the barrel within which anyone can fall over the edge of corruption, embezzlement, and even sexual abuse? A few months ago, I caught myself on the edge of corruption. Arriving to Rwanda without my office having arranged a Congolese visa for my foreseen three weeks stay in Bukavu, I immediately panicked at the idea of not being able to complete my mission there, anticipating a bad review from our project evaluators and a disappointed donor at a simple logistical mistake. I entertained the idea of attempting to bribe a border official to enter the country. I thought that to carry out my "peace"-related mission, the ends would justify the means. Understanding peace as an end, an objective, an outcome, is exactly what has precipitated many others to venture over the edge.

Using decolonizing research methods and critical pedagogy, this chapter will question the epistemological structure of peace as we know, practice, and teach it, as well as the implications of our thoughts and actions on the daily lives of the populations we are supposed to serve. By way of an illustration, I will take a close look at the resurgence of human trafficking in post-conflict areas, and will question how and why, in spite of an array of international legal instruments, internal UN policies and lessons learned, it is still one of the most widely practiced forms of abuse worldwide. This will lead me to question the epistemological basis of peace, as a theory, a practice, and a discipline. Finally, this chapter will make the case for a different approach to peace, one that does not rely on "benevolence" or any other colonial narrative that serves the social and economic interests of a complacent ruling elite. Decolonizing peace calls for an introspection of all aspects of the peace industry, the transcending of a structural elite toward the formation and facilitation of endogenously sustainable communities of peace processes. It brings parts of the invisible to the forefront. It involves the dismantling of "official" narratives, asserting the first person and subjective experiences of all those involved as visible and relevant.[17]

Huda and Sajeeda

In the Spring of 2003, as the US-led coalition's infamous de-Baathification program dismantled all law enforcement, legislative, and military Iraqi institutions, organized crime in Iraq was given a free hand to resume its pre-war activities, and extended its scope.[18] Sisters-in-law Huda and Sajeeda were abducted by armed men on a September morning while they were cleaning their front porch.[19] They were drugged, beaten into submission, and sold off to a pimp. A few days after their abduction, they were given fake passports and driven with their new "owner" through the Syrian-Iraqi border. Since borders were not protected by the coalition, not being a strategic priority in the "building" of a new Iraq, they were unable to alert any official at any point.[20] Upon reaching Damascus airport, they mistakenly thought that customs officials were going to help them. They pleaded for help to whoever was willing to listen to them. Their hope for salvation was crushed after money was exchanged between their pimp and officials, and they were subjected to a severe beating for trying to escape. Upon reaching Yemen, they started working in a hotel with another 180 Iraqi women and girls. The youngest among them was 11 years old. After a few weeks, they managed to contact their mother, and mother-in-law, Aisha, and asked her to organize their rescue. Aisha went to the Iraqi authorities, to no avail. She then tried the coalition, and was given a sympathetic ear by a US sergeant on duty. While he could not enforce any legal provision to have them freed, he assisted her in putting pressure on the Yemeni embassy for their police to take action. By then, I had alerted Amnesty International of their case as well. This combination of efforts led to their hotel being raided by the Yemeni police in April 2004. All women were put in buses, and taken to Sana'a airport. They thought that this was the end of their ordeal, only to realize that they were left at the airport with no passports or money, trapped in a country whose authorities, or well-intentioned international NGOs, were unable and unwilling to offer any assistance outside of their budgetary scope. Most women settled for having their pimp marry them off, for a large sum of money, with the hope of returning to Iraq at a later date. Others made a deal according to which they would return to Iraq and work for their pimp in a brothel. Huda and Sajeeda were among these. As soon as they reached Baghdad, they escaped and returned home. While Huda's parents welcomed her with open arms, Sajeeda was threatened to death by her brother if she did not accept to divorce her husband and return to the family home, to be kept there under lock and key for the rest of her life, as her abduction alone was thought to have tarnished her family honor. She disappeared shortly after her return home and has not been seen since.

As far as Amnesty International was concerned, it had done its job by raising awareness of a heart-breaking issue. After all, there was no line on their sophisticated, London-elaborated budget for the repatriation of trafficked persons or their protection if/once they had returned home. Maybe a string of other agencies ought to have picked up on it, but a lack of co-ordination made this impossible. Still, this story, at the tip of the human trafficking iceberg, could have been carefully packaged to ensure a continuous flow of individual donations to their London-based office.

Moreover, while organizations such as Amnesty International have never made their yearly operational costs thoroughly transparent, the percentage of donations that actually benefit people on the grounds of their awareness campaigns is thought to be less than 20% at best.[21]

Human Trafficking and International Law

While commiserating on the desertion of Huda, Sajeeda, and their companions, one may feel consoled to know that International Law does protect what it refers to as "victims" of human trafficking. The ordeal of Huda and Sajeeda falls within the remit of the Palermo "Trafficking

Protocol" of 2000, as part of the Convention against Transnational Organized Crime.[22] Human trafficking is defined as:

> *42(a) the recruitment, transportation, transfer, harboring or receipt of persons, by means of the threat or use of force or other forms of coercion, of abduction, of fraud, of deception, of the abuse of power or of a position of vulnerability or of the giving or receiving of payments or benefits to achieve the consent of a person having control over another person, for the purpose of exploitation. Exploitation shall include, at a minimum, the exploitation of the prostitution of others or other forms of sexual exploitation, forced labour or services, slavery or practices similar to slavery, servitude or the removal of organs;*
>
> *42(b) The consent of a victim of trafficking in persons to the intended exploitation set forth in subparagraph (a) of this article shall be irrelevant where any of the means set forth in subparagraph (a) have been used.*[23]

While they were trafficked out of Iraq, Huda and Sajeeda were supposed to be protected by the 1st Protocol of the Geneva Convention, Article 75, as they were to be the "object of special respect and [...] protected in particular against rape, forced prostitution and any other form of indecent assault."[24] Even though the US government never ratified the convention, it was and is still supposed to abide by it according to customary law when it is occupying a country.

Since full sovereignty was handed over to the Iraqi Transitional Government on June 28th, 2004, the case of Huda and Sajeeda fell under the remit of the aforementioned Geneva Convention. Yet they were let down at many stages during their ordeal. The borders that they were forced to cross were not safe. They were not helped by the Iraqi police anti-trafficking unit, which, when I interviewed them, had decided that they had eloped with a pair of "lovers".[25] Their mother, and mother in law, Aisha, did not benefit from any institutionalized structure within the US-led coalition to help find her daughters. Instead, she relied on the good will and heart of a US soldier, who could equally have turned her away when she came to him for help. When they returned, Huda and Sajeeda were not afforded the protection of a safe house, where they could stay until their families decided to take them in. Rather, it is more than likely that Sajeeda either returned to the claws of sexual slavery, or was the victim of an honor killing. To this day, she has not reappeared. International law provides a hopeful response to the issue of human trafficking, yet in effect, fails to ensure the safety of human beings. From this perspective, we can see the wisdom of linguist Alfred Korzybski's famous phrase: 'the map is not the territory.'[26]

Abu Baker, Buk and Koran

Human trafficking is not only limited to sexual slavery, it can also take the form of forced domestic labor, forced labor, reproductive slavery, etc.[27] According to a UN estimate, human trafficking touches 2.5 million people worldwide, and the annual profit it generates amounts to approximately US$ 31.6 billion.[28] In post-Saddam Iraq, human trafficking has recently reached a new height, in the form of forced domestic labor. Brought in with democracy and peace is a new fashion among the Iraqi elite Middle Class, the "ownership" of a house slave, or two. Should a shop be successful, having a Bangladeshi serving its customers is considered the height of refinement and the envy of one's neighbors. Asian migrant workers now fill the streets of Iraqi Kurdistan in Northern Iraq, as street sweepers, garbage collectors, painters, laborers, etc. They have also started appearing as domestic laborers for NGO workers and the international media. Meet Abu Baker, another piece of Korzybski's "territory." Abu Baker is a Bangladeshi Sunni Muslim migrant who works in a hotel rented by one of the leading news wire agencies in the western world.

For the last year, he has been working between 12 and 14 hours per day, cleaning the 12 rooms of this small hotel, and handling the guests' laundry and cooking, when possible. He came in from Saudi Arabia to Iraq, as he was told that working conditions in Baghdad were better, and since his arrival, his passport has been withheld. He is supposed to be paid $200 per month, but since he is expected to reimburse his employer for the airfare that brought him to Iraq, he never gets to see his wages. He is constantly hungry, sleeps very little, has no vacations or days off, no health-care, and sleeps on the hotel's kitchen floor.[29] In 2011, Iraqi legislation was enacted to "protect" people like Abu Baker. It has decided to no longer grant visas for migrant workers, although it is going back and forth on its decision.[30] Will this decision have any impact on his daily life? Will it help him at all?

Next door to Abu Baker's hotel are Buk and Koran, two migrants from Nepal, working in a house occupied by an Iraqi security company. They each earn $500 per month, $300 of which is transferred directly to a bank account in Nepal.[31] They assist one another in the villa's daily chores. They have a separate living area outside the kitchen, with bunk beds. They eat as much as they need and have one day off per week. Their employer did not give in to their recruitment agency's request to have their airfare reimbursed, so their salaries are not "taxed" for reimbursement each month through debt bondage. Their passports are in their possession, and until the government's decision to no longer grant visas for migrant workers, they thought that they would benefit from a free return ticket home for 10 days of vacations per year. While the Iraqi government is going back and forth on this decision, should it be maintained, how will it affect both them and Abu Baker? This depends only on the good will of their employer. Given the slavery situation in which Abu Baker finds himself, it is likely that his employer will force him to work until he no longer can, either falling gravely ill or dying on the job. For Buk and Koran, this means that they will be compelled to work without a vacation until they decide to go home for good.

In both cases, workers are trapped in Iraq, to different degrees. When many workers like Buk and Koran decide to return home, the "pool" of available workers in the country will decrease, opening the door to further abuse from employers. It is likely that "workers" will become more expensive to "acquire", and that "visas" will have to be paid for handsomely, resulting in a higher risk of debt bondage. For instance, since there will be less availability of new "blood", it is likely that "owners" will seek to keep their staff at any cost, preventing them from having any contact with the outside world which might lead them to better employment opportunities elsewhere. At the time of writing, the Iraqi government has indicated that it may lift its visa ban and begin "charging" between $500 and $1000 per visa. This amount would undoubtedly be re-paid by workers on the long run.[32]

The foreign journalists that live in Abu Baker's hotel do not seem to take notice of the living conditions of the man serving them on a daily basis. They are busy completing their daily tasks, championing press freedom and Iraq's newfound democracy, despite staying in a hotel where freedom only exists for the chosen few. This contradiction is worth more than mentioning. It forces us to acknowledge the fragmentation and limitations existing within post-conflict settings, whose deeply flawed peace paradigm only caters for a fraction of the populations is it supposed to reach. Within this paradigm, there is a hierarchy of human beings, values, origins, and ethnicity. This paradigm values the map above the territory, and, through a familiar religious narrative, rewards the chosen few for their best behavior, while promising the wretched that there is a place ready for them, peace heaven, should they behave in a way that will not upset the current social order. Abu Baker, be consoled, for you will be rewarded a hundred times when justice finally breaks out alongside sustainable peace, on judgment day. A few peace and conflict scholars have chosen to define this paradigm as "liberal peace", the linear, mechanistic building of peace as an aggregation of parts built through a liberal framework.

The Liberal Peace Paradigm

The critique of liberal peace questions the fragmentations, inconsistencies, and priorities regimenting post-conflict environments. It asks why some individuals are worth more than others, and who benefits from a peace whose expression is as industrial as its promotion is idealistic. A probing of liberal peace asks why Huda, Sajeeda and Abu Baker are left behind while billions of dollars are poured into post-conflict areas on the basis of their "vulnerability". Oliver P. Richmond in *The Transformation of Peace* has provided reflections on the liberal peace model, which he understands as the universal, neo-colonial, state-building model applied indiscriminately in post-conflict missions after the Cold War. According to this model, salvation and sustainable peace in post-conflict situations are based on the construction of state mechanisms through the promotion of good governance, free market, law enforcement institutions, and human rights.[33]

The neo-colonial flavor associated with this enterprise relates to the idea that Northern-educated, seasoned "democrats" will be deployed to educate locals on the values that they ought to embrace and be grateful for. An illustration of this would be my own story: the deployment of an inexperienced, 25 year old French democratization officer to a small village in Bosnia Herzegovina, where only her passport credentials, her nationality, and birth-rights, endowed her with the privilege to "empower" women, youth, and politicians often twice her age, with no prior experience. This situation, Richmond suggests, often leads to the local rejection of or resistance to perceived neo-colonial institutions and models, as well as the resurgence of conflicts in many parts of the world.

Of importance to liberal peace is the fragmentation and prioritization of certain domains over others. Often, these priorities are dictated by the Northern capitals that these state-building enterprises emanate from, and at times create some extreme examples of the disconnect between what is considered policy priorities and some of the values these missions are supposed to champion or protect. An illustration of this would be the latest scandal involving Washington DC-based security company Dyncorps, regarding a cable between an Afghan and a US diplomat who discuss the purchase of dancing boys, *Bacha Bazi*, for Afghan policemen in exchange for their assiduous participation in a police-training program.[34] In the early 2000s, this same company was also involved in a human trafficking scandal in Bosnia Herzegovina, where it was supporting the United Nations International Police Task-Force.[35] Over a period of months, a Dyncorps employee collected information on the buying, selling, rape, and murder of women as well as children as young as 12 years of age.[36]

To this day, Dyncorps continues to be contracted by the US Department of state in Iraq, Afghanistan, Sudan, and other places where the liberal peace paradigm is being applied.[37] In light of this, Mary Anderson's *Do No Harm* essay raises the question of a Hippocratic Oath for peace missions personnel, since, as international actors in times of conflict do become part of the conflict itself, an epistemological distance is called for.[38] Is this enough? More importantly, are peace workers aware of the contradictions and shortcomings that their presence represents? A Hippocratic Oath places the responsibility of success onto the shoulders of peace workers, while a critique of the liberal peace paradigm implies that the system itself is bound to fail. Peace mission personnel are not benevolent physicians at the side of a sick nation, society, etc. This narrative encourages challenges to peace to be seen as a sickness, an ailment to be cured, and in so doing may actually promote the renewal of an elitist system complacent to a neo-colonial power structure. Is the system bound to fail or re-create the same social order? Development efforts and their shortcomings are a constant reminder of this conundrum.

Indeed, Dambysa Moyo and William Easterly have deconstructed the logistical arm of liberal peace, development, and international assistance with great eloquence. In her book, *Dead Aid*, Moyo describes the systematic dependency of African nations toward the industry of international aid.[39] She contends that aid as a development model cannot function or be sustainable in the long term, as it creates a pattern of reliance on foreign economies and upholds the power structures of local elites. From her perspective, there is no incentive for sustainability in relation to aid, since it creates jobs in the North, and maintains a socio-political status quo in the South. In *The White Man's Burden,* William Easterly articulates the perspective that aid is not only an industry, but also a neo-colonial arm of Northern powers working to keep the greater South in a cycle of dependency and despair, while simultaneously granting themselves a moral self-righteousness for helping the world's poor and pulling them out of "darkness."[40] His main argument condemns the vision of planning, or "planners" based in Northern capitals, masterminding operations to save Africans from themselves, while seldom understanding what needs, realities, and initiatives are endogenous to those environments. Easterly advocates for the morphing of planners into "searchers", who already do exist, and instead of applying one-size-fits-all solutions to Africa's woes, identify and customize actions according to local realities. Both Moyo and Easterly emphasize the fact that the billions of dollars spent to supposedly alleviate poverty seldom reach the people that desperately need it, and instead feed the egos and budgets of international NGOs and their idealist, Ivy-League-trained, "white" aid-workers.[41] Scores of articles have been written on the legitimate channeling of funds that make the aid community a business and an industry. Two illustrations come to mind.

The first one refers to a USAID-sponsored project I was asked to evaluate and monitor in 2005.[42] This project was designed to organize the first three elections of the new Iraqi electoral cycle from a logistical and voter outreach perspective, and to support the drafting of the provisional Iraqi constitution. Out of the $155 million allocated by USAID for this project, an estimated $100 million was directly allocated to contracting companies for their services, primarily to ensure the security of the aid workers involved in these various projects.[43] The actual funding delivered to communities for voter outreach or electoral document circulation was a fraction of this initial budget. Equally, the funds allocated to our evaluation project were minimally allocated to deploying field monitors around Iraq to assess the impact of the programs initiated by the USAID partners. Since we were not allowed to leave the safety of the International Zone inside Baghdad, more commonly known as the Green Zone, it made sense to allocate a strong budget for field research. This was not the case. Almost every dollar, from our transportation exclusively on American carriers to our daily spending at the local Subway franchise of the Green Zone, channeled our budget, spending money/per Diem, and salary back to Northern-based economic interests. This particular case presents more than an anecdote, it demonstrates the spending re-cycling of aid money, rarely channeled outside Northern business or NGO interests.[44]

The second illustration is an NGO I visited in Congo (DRC). The organization had been working to help rape survivors for the last 10 years. It then realized that, throughout this period, less than 20% of the funding raised in Belgium, where its Headquarters are based, actually reached the women it was supposed to help, despite following all applicable laws. Is peace as much an industry as any other? Is "peace" geared toward giving our Northern peace and conflict studies graduates careers *and* a good conscience?

The Field beyond Right and Wrong

Common to all these critiques of the peace-building world, whether it pertains to the discourse of liberal peace, international aid, or the planners' perspective, is a narrative of responsibility to save the greater South from itself, to bring it to "our" Northern level and standards – economically, politically, culturally, and legally.[45] The tools: democracy, state-building, good governance, transparency, accountability, human rights, and rule of law, are often lagging behind in our own Northern environments, i.e. Guantanamo Bay Detention Centre, Oil for Food scandal, etc., yet are presented to our Southern Other as their salvation, totems to the altar of their forced modernization.[46] In fact, as pointed out by Easterly, the past rhetoric of colonization vividly matches that of the main United Nations documents today.[47] The liberal evangelization of the 21st century is similar, in its narrative, to its religious counterpart of previous centuries.

This realization is what makes me, as a "practitioner", question the foundations of my discipline. What am I to say to my students from the global South when they deplore the fact that most of the theories we teach emanate from the greater North, using them, the South, the "Other", as mere case studies, as the problem to be solved? What am I to answer to well-intentioned colleagues when they come back from years of dedication to the "field" burned out and disillusioned? What advice am I to give my students who ache to go to the "field" to make a difference in the world? With "field", of course, comes the Orientalist characterization of the "Other" from the greater South.[48] In the same way that evangelization presented the glorious map of heavenly life as an ultimate reward, the liberal peace evangelists present the map of a magnificently just social order to the populations they so benevolently "help", while quietly earning money for it and maintaining a social status quo within the territory.

Sufi poet Rumi wrote: "Out beyond ideas of wrong-doing and right-doing, there is a field. I'll meet you there."[49] I do not want this reflection to take one side in relation to the impact, or not, of peace missions: other colleagues have already carried this out beautifully.[50] Rather, I wish for it to transcend a too obvious polarization that may have been generated by alternative voices in the field. I am aware of the dedication of my peers, of the positive impact of some projects on the ground, of the hope that some of our actions might bring: "success", "development", "relief", or at least a modicum of comfort and kindness in an overwhelmingly uncomfortable and unkind world, both to "us", and also "them". I also know that the road to peace mission hell, embodied by the resurgence of the Afghan *Bacha Bazi*, is paved with good intentions, and materialized by the rule of law, in this case, Afghan police reform. What I am challenging here is the assumption that the liberal peace model is what the "Other" needs, the assumption that any model that suits our good intentions is needed. I want the lessons learned to matter, so that children, women, and men, no longer suffer the consequences of our recurring mistakes. One does not need to break lives to make peace, for it then only represents a victor's peace. Finally, I also want to question the commonly expressed assumption that malpractices are exceptions to the norm. What is it, within the system, which transforms the exception into the norm?

No solutions

At the University for Peace (UPEACE), I co-teach the foundation course in peace and conflict studies. I have been doing so since I arrived at the institution, and the content of my teaching, as well as my methodology, has changed drastically in the last few years. The main factor responsible for my different approach has been the diversity of our student population. Teaching peace and conflict studies in a Northern institution, where a strong percentage of the student body is homogenous, does not challenge the system within which one evolves at all. After all, this teaching will endow its privileged elite to go and save the world after they graduate. When I arrived at UPEACE, the base of my teaching was peace and conflict studies classics.[51] After a couple of years repeating the same models, I became increasingly aware that the majority of our students, from the greater South, were weary of both the theories and case studies utilized in the course. As mentioned earlier, they were being "studied" with a Northern eye. Some felt that they were being objectified, while others questioned their place within a university setting where they, the majority, perceived themselves as a minority. Were the conflict resolution mechanisms that we were teaching applicable to individual or collective settings? Where was the place for alternative conflict resolution mechanisms outside the rhetoric of "indigenous" traditions? Where does the universal finish and the indigenous start? Why would the Rwandan *Gacaca* practice, or the Hawai'ian *Ho'ponopono* process, be considered by our literature as "indigenous", while Fisher and Ury be understood as universal?[52] At the same time, questioning the liberal peace paradigm raised some frustrations within parts of the Northern student population. The critique of liberal peace, it seems, was challenging the dreams and also the forming identity of some.

One specific interaction embodied this conundrum. After my first lecture, which deconstructed some of the Northern assumptions of peace and conflict studies, one British student was particularly upset. He emphatically let me know that just about every argument of my lecture had profoundly irritated him. His reaction was understandable: he was fresh out of his undergraduate studies, with little practical experience. He was too young in our "field" to be critical of the liberal peace narrative. At the same time, he was also enrolled in a graduate school where he ought to have expected to be intellectually challenged.[53] A deconstruction of the liberal peace paradigm had just touched his idealism at its core. If liberal peace was the expression of the status quo, and the aid paradigm an industry geared to maintain this status quo, where were his good intentions going to fit? In a meeting subsequently held, he expressed that he was reconsidering his decision to stay at UPEACE. He said that he had just spent an enormous amount of money to be given the tools that he would need to make the world a better place. Being told that there was no universal tool to alleviate the world suffering was just too much for him to take. He asked if he would be spending the rest of the year in his particular program, International Peace Studies, deconstructing liberal peace. I replied that at the beginning of the year, yes, we would deconstruct a great deal of theories and practices. However, after this initial phase, we were also going to explore some pathways toward post-liberal peace, decolonizing peace, as well as to offer alternative proposals to this conflicting paradigm. I also reminded him that, no, I had no universal tools to propose to him. In fact, a transcending of the universalism of liberal peace was the only certainty that I could propose to him. Peace, in the program, was going to be seen primarily as a process. We were not going to make a "peacebuilder" out of him. No community is an empty shell, nor does it need a fresh graduate to come and "build" peace from scratch. Rather, we were going to train him to facilitate peace as a process, if anything. Still, if he insisted on being labeled, he would leave the program as a peace-facilitator, or a vector of peace formation.[54] This, I contended, was the only intellectually honest label that I had to propose. A few days later, he switched to our law program. The map had more appeal than the territory. He now works for NATO.

Liberal Peace and its Enabling Great Caveat

The tools, little handbooks on, steps to, and paraphernalia of liberal peace, are legion. Peace, on paper, is a very straightforward matter. With a careful mix of good governance, rule of law, accountability, democracy, transparency, and multi-track diplomacy, a country can be sustainably transformed for the better. In fact, after reading these books, one with no field experience might get the impression that peace-building is like assembling a car, and that fixing a country and its people is a viable possibility. Since the 1960s, the greatest caveat, the safety net of liberal peace, has existed in the form of a false dichotomy between negative and positive peace.[55] We are told, and are still teaching, that, yes, harmonious peace is difficult to achieve, and that while negative peace, the absence of war, can be engineered mechanically, positive peace, or the fairy tale of happily-ever-after co-existence, remains complicated to achieve, yet attainable. We are told that positive peace takes years to foster, that it is no easy task, but that our good intentions, in the end, will prevail. While useful in its time, the cob-web-made safety net of positive peace still absolves us from questioning our entire paradigm. It enables us to retreat behind its safety when everything else fails, in the same way that Dyncorps' repeated scandals will be understood to be the exception to the rule. How often have we heard that, after trying everything, it was just impossible, within this or that failed state, this corrupt political culture, or amidst that reality of vicious/atavistic ethnic hatred, to do anything more for "those" people, the "Other", who are not like us, do not understand our peaceful values?[56] How dishonest can we be to even assume that our mechanical actions bear no responsibility in the situations that we too often flee, evacuate from, or abandon?[57] In every such situation, there are people, civilians, including children, who will not survive our liberal peace paradigm, who will die because of it.[58] Regardless of the relevance of positive peace at the time, its obsolescence now makes it the enabler of our failures. We cannot but transcend it.

The Peace Universalism

Where does this paradigm emanate from? We may all agree that while we look into the transitions between stages of peacekeeping, peace making, and peace building, we are not looking at a strict linear progression. We may also be aware of the complexities existing within these classifications. For instance, we know that within one particular case, each of the stages mentioned above might be present at the same time. One case very obvious to me in relation to this is the situation of post-Saddam Iraq between 2004 and 2006. During this time, while a democratization process was taking place, generally classified as "peace building", many parts of the country were experiencing acts of ethnic cleansing, more related to the sphere of peace keeping.[59] The merging of all occurrences, and their complexities, can be seen to account for the failure of liberal peace in post-Saddam Iraq. We are still telling our student population that these models are the best we can offer them. Indeed they are. These models emanate from a paradigm that has been dominating our thinking for hundreds of years: Cartesian thinking.

In *The Turning Point*, Fritjof Capra analyses the key evolutions of Western/European thinking between the years 1500 and 1700.[60] While before 1500, the dominant view of the world combined both mind and matter in organic, collectivist and ethics-based communities, the arrival of the Scientific Revolution by way of the seminal works of Copernicus, Galileo, and Descartes, privileged a vision of the world as a machine to be tamed, controlled and engineered.[61] This vision of the world as a machine prevails today in our daily lives, e.g. the separation of academic departments within one university structure, or the prioritization of specific objectives within peace missions, at times privileging state building over basic human rights, as was the case with the lack of interest in countering human trafficking in 2003-2005 post-Saddam Iraq. While the Scientific Revolution was extremely useful at the time, and still is now, the separation of humanity from its environment, spirituality, and ethics, has brought a lack of equilibrium that is now culminating in, to name only a few impending catastrophes, climate change and nuclear disaster.

In *The Structure of Scientific Revolutions*, Thomas Kuhn argues that scientific evolution is not the result of a gradual process, but of revolutionary changes affecting the way we conceptualize the world as a whole, our social paradigm.[62] The Cartesian revolution, separating mind and matter, did transform our daily lives into compartmentalized settings, where one wears different hats at different times of the day, and where one's fragmented vision of the world transpires in one's actions. It is thus possible for a person to work in a peace mission during the day, and have a sex slave waiting for him/her to return home at night.[63] True, we can all reach post-conflict environments thanks to the technological advances of modern science. Yet, it is also modernity at its core, in its sheer methodical ruthlessness, which engineered and executed the Jewish Holocaust, validated colonization, and now fuels neo-colonialism.[64] At stake in this debate is not whether modernity is needed or not, rather, it is how the reliance on hard sciences as a paradigm, a model on which our lives are based, have brought disequilibrium to humanity. The word "humanity" is a very conscious choice as applied to this text. While the Cartesian thinking of "I think therefore I am" emanates from Europe, it is the paradigm that it has created that has permeated throughout the world, culminating in globalization and the universalism of peace studies.

Thus, when I see that to be a successful academic in an African University, one has to be educated in a Northern-based institution, and return as the prodigal son or daughter with these unfaltering credentials, I also deplore the pervasiveness of this neo-colonial model. Where is the space for academic creativity and innovation, if there is only one paradigm that dominates the thinking of elites worldwide? Recently, I was assisting in the development of a syllabus on good governance for a University in Eastern Congo. My colleague, a brilliant and resourceful professor, gave me an appalling first draft of his curriculum, listing governance models of the European Union, the World Bank, United Nations, etc. When I asked him how this would fit in Eastern Congo, he realized that his curriculum could be offered as part of any generic program anywhere in the world. When I asked how his region fared, was organized, and administered itself before colonization, he wondered if he could base his work on a traditional governance model to be applied to today's realities, mitigating all influences to fit his environment.

We are not calling for a return to the basics, to the proverbial "cave", but are merely looking into what governance could possibly mean in a contemporary Congolese context. Are some lessons learned from the past useful for our future? To date, very few models supersede the "indigenous" label that the imperialist Northern academia grants any of its competitors. Cartesian thinking as a paradigm, indeed, emanated from the West, and condemned all other paradigms to the "indigenous" label. This also alludes to the universalism of the liberal peace paradigm. Not only is peace administered as a mechanical remedy to conflict, its main precepts are also understood to be universal.

Decolonizing Peace

The case for decolonizing peace comes from the realization that the same paradigm that was invoked for colonization is now serving to channel neo-colonial liberal peace efforts worldwide. Decolonizing peace calls for a holistic, systemic approach to peace, the processes that represent it, and the ethics and values that it embodies. Decolonizing peace means harmonizing the map and the territory, bringing the invisible to our understanding and living of the visible. It calls for mitigation between localized social fabrics and values of peace, it also questions the idea of imposed change at any cost, usually that of a peaceful process. Contrary to colonial discourses, which once invoked that Africa as a whole was there to be discovered, conquered, and built from scratch, the narrative of decolonizing peace asserts that peace already exists at the local level, and that it does not need to be built according to values and understandings of a foreign environment.

Is there universality in ethics, human rights, and a culture of peace? Decolonizing peace does not call for a discarding of human rights or values of peace for the sake of cultural relativism. It does not call for a return to a basic order of patriarchal exploitation, for instance.[65] It questions the political and neo-colonial motives that are being pushed through the championing of certain "universal" values. When the map is no longer pitted against the territory, sustainability is not to be engineered, doctored, or re-invented: it is intrinsic to decolonizing peace. The role of the "peacebuilder" comes to be heavily challenged under the guise of decolonizing peace. The Northern-white-ivy-league educated expert and its elite-Southern-born-Northern-educated counterpart are asked to reflect on the compartmentalization of what they have understood to be an end justifying all means.

Decolonizing peace requires a paradigm shift that enables its practitioner to see initiatives through a different set of lenses and to employ a different array of what are no longer "tools" but understandings of what can be facilitated, strengthened, and enabled to flourish on the "ground".

This paradigm shift transcends the power dimensions of social orders, it also evolves outside of the left-right political spectrum, for politics also kill the sustainability of any decolonized initiatives, as Nobel Prize winner and Indigenous leader Rigoberta Menchu's case would illustrate.[66] Of importance is the idea that peace becomes a decolonizing agent, but not a decolonized finality. It evolves within the complexity of a live adaptive system; it is always in motion, as a process. It goes through cycles, vanishes to re-emerge in another form, according to the bifurcation point that it takes through its constant adaptation and re-invention. It cannot be controlled, since it is a complex adaptive system.

A Different Lens

Wishful thinking or parallel reality; what does decolonizing peace look like? The paradigm shift that decolonizing peace represents is also in the eye of the beholder. It is not for this book, therefore, to convince anyone of its relevance. As mentioned earlier, an internationally funded NGO focusing on rape survivors realized that its activities were mainly benefiting a handful of elite employees both in Belgium and DR Congo. As less than 20% of its donations from Belgium actually reached rape survivors, they decided to re-evaluate their liberal peace activities of capacity building, women and youth empowerment, legal aid, and financial support to hospitals, to strengthening local initiatives.

The issue of rape in the DRC has been prominent in the international scene for some years. Thousands of women have been helped as a result of numerous campaigns. They have gained access to much needed surgeries, medication, and psychological assistance. In some cases, they have been given a new lease on life, and the issue of rape in war has gained much more exposure internationally. None of this much-needed assistance deserves to be questioned, yet, is it sustainable on the long run? There is one town in the Southern Kivu region of DRC where all this assistance is gathered: Panzi. Due to the concentrated media and academic coverage of this particular town, almost 100% of all medical services given to female rape survivors are provided there, neglecting the development of similarly relevant infrastructures elsewhere in the region.[67] Moreover, most reports focus on female victims, neglecting the equal prominence of sexual violence targeting men, hence the unavailability of treatment for male rape survivors.

Taking these parameters into account, alongside many more, the Belgian NGO decided to alter all its activities two years ago. It changed its local staff into a locally educated, gender-equal, and ethnically diverse team, and suspended all its liberal peace programs. It now focuses on logistically and financially supporting diverse homegrown initiatives and NGOs working on minimal, locally gathered funding, solely with local staff. Its headquarters have been transformed into a half-way house for survivors and local NGO workers to meet and work toward medical assistance, legal help, small business initiatives training, and very importantly, the emerging of a network of survivors to help one another once they return to their home environments.

The organization is also helping local hospitals and clinics so that survivors do not need to travel all the way to Panzi to benefit from medical treatment, thus provoking less of a negative impact both socially and family-wise. The treatment of men remains an issue, but has become part of the organization's agenda. No international "expert" has been detached from the Belgium headquarters to decide what its priorities ought to be. Rather, a local manager reports to Belgium. The idea behind this thinking is that the existence of local NGOs cannot depend on international money. Rather, those NGOs are based on a strong social fabric of mutual trust, resilience, belonging, and values of care. Money certainly helps for those values to blossom, but could never engineer them. A different set of lenses on part of this Belgium organization, coming from a paradigm shift regarding their own role and standing, made a difference to a resilient community, and fostered its sustainability. This bears the hallmarks of a decolonizing process, where an equal space exists for all.

Decolonizing Methodologies: Teaching and Writing to Transgress

It will not have escaped you, the reader, that this book is written with a semi-autobiographical tone, also placing other individuals at the forefront of the overall narrative. We are often told that individuals cannot be singled out, exposed, because they need to be "protected." Those individuals are moved from being seen in their humanity to being labeled as objects of our "independent" ethical research.[68] All individuals portrayed in this book except Mrs. C. pleaded with me to make their lives visible. From their perspective, they do not need "protecting": they want their ordeals to illustrate our failures as benevolent "peace builders." They seek to make us accountable in light of our own fiascoes. They want to matter. How can I, as a peace and conflict studies professor, training a little army of "peace builders" detach myself from the failures and crimes of our peace industry? An autobiographical style is also a statement of my own responsibility as a practitioner. It may be seen as a transgression from more conventional academic writing, yet it is a chosen stylistic transgression toward empowerment, awareness, and action.

The choice of the territory over the map stems from the epistemology of decolonizing peace. The individual is no longer portrayed as the problem, the exception to the beautiful rule on paper; rather, he or she represents the call for accountability in peace missions, theory, and teaching. The individual warrants the decolonizing of all, a reappraisal of peace as a process rather than a mere end. This also stems from the choice to teach and research peace and conflict studies from the perspective of critical and engaged pedagogy.[69] In teaching for and through decolonizing peace, there can be no boundary between the "subject" and "object", no barrier separating the researcher and the researched. In terms of research methods, all "material" for this book was gathered using Linda Tuhiwai Smith's decolonizing methodologies.[70] Stemming from the realization that conventional research methods can no longer be seen as an independent, clinical tool for data collection, a decolonizing methodology re-appropriates control over the formation of knowledge and one's understanding of their own reality. It becomes a dialogical process between all involved, stems from being more than participating, and does not rely on pre-established keywords and categories. Teaching, researching, and writing to transgress liberal peace becomes the process of decolonizing peace.

Conclusion

As I reflect on decolonizing peace, one of my former students has just been hired as a protection officer for an NGO in South Sudan. She is street-smart, under 25 years old, and full of ideals and energy. She has been working with me on this emerging paradigm for some time. Still, I wonder how many of my students will eventually be driven over the edge by the liberal peace system in the long run. As I set off to continue researching the abuse of local populations in DR Congo by MINUSCO staff, I hope to also remain in this parallel universe, not to be coerced into accepting the unacceptable.

Amidst former allegations of human rights abuses and sexual slavery, the UN mission to DRC was re-named as MINUSCO in 2010.[71] All controversy has been relegated to the past, with the assault, rape, and enslavement of women and children by UN staff and soldiers no longer tolerated as part of this new and improved mission and mandate. True, appearances are slightly more presentable. UN vehicles are no longer parked in front of nightclubs in and around Bukavu. When they do not use civilian cars, peacekeepers now send in their *petits*, errand boys, to bring them the children that they will safely rape in the confines of their environments.[72] Then they will build a mosque or a church to make it all even, to thank the local communities for having given them their pound of flesh. All is a little more difficult to track down, but still very real.

Where does this leave us as "practitioners"? Huda, Sajeeda and Abu Baker remain at the mercy of our good conscience, our willingness to search our own souls. Still, we can no longer say that there is no other way to approach peace. A parallel world exists: we have to acknowledge it and evolve according to its needs, even if this makes a part of us redundant.

II. A Leap of Faith

How many are we who try to make a difference in the world, only to end up deconstructing its imperfections to the infinite? Do the contradictions of the peace industry exposed in Chapter I warrant a rejection of peace as a concept, or the abandonment of our hope for peace? After being committed to helping others, to being the peace they wanted to see in the world, some have left the field of peace and conflict studies altogether, opting for careers in finance, real estate, or the film industry. Does the realization that peace has transformed itself into an industry merely warrant one's cynicism, or, to the contrary, does it call for an even greater commitment to its values and principles?

Decolonizing peace stems partly from a decolonization of the mind; from the cognitive and emotional understanding that individuals do not necessarily need expert outsiders and their resources to shape their daily lives, or more importantly, to bring them peace. In exploring Rumi's field beyond right and wrong, above the Cartesian duality of false social choices, a slice of life from El Salvador will allow us to analyze some of the everyday dynamics of the peace-excluded middle, as well as their backlash. By way of exploring a paradigm shift, this chapter will review certain dynamics of social change and recurring dimensions of inequalities and hierarchies. It will then analyze how the discursive and structural fundamentals of peace as immanent may nurture the neo-colonial expression of a peace to be built by outsiders, and not by affected individuals and their communities. A last part will look into some ground realities and consequences of the immanence of peace, as a basis for opening one's eyes to alternatives that are already taking place worldwide. This chapter will conceptualize a paradigm shift, from peace as transcendent to peace as imminent, as a leap of faith.

"If God Existed"

Geovanni Morales is from the Mejicanos district of San Salvador, one of the most deprived neighborhoods of the city.[73] There is not one week that goes by without scores of casualties in the war between its gangs, the Mara Salvatrucha (MS-13) and the Mara 18, reinforced through drug trafficking, organized crime activities, and police brutality, all within a machismo subculture of hegemonic masculinity. In spite of the common outsider belief that the daily violence in El Salvador is the result of the gang phenomenon, inhabitants of Mejicanos assert that those gangs would never have emerged if it wasn't for the misery, social injustice, state repression, and the civil war that displaced and scattered many families between 1980 and 1992. It was at the age of nine that Geovanni joined the MS-13, in his own words, to have a family and a roof over his head. One day, as he returned home from an errand, his mother was gone, as was the rest of his family. He stayed with a neighbor for some time, waiting for anyone to come home, and when he had outstayed his welcome, his only option was to join the MS-13. It was either this, or live under a bridge until he grew old enough to be forcibly enlisted in the Salvadorian army to fight the Frente Farabundo Martí para la Liberación Nacional (FMLN) rebels. To this day, he does not know whether his mother is dead or alive. For years, he was an active member of the MS-13, involved in drugs trafficking, petty crimes and street fights. According to him, the life he was leaving was not particularly to his liking, since he refers to himself as way too sensitive for the machismo culture that prevails within the gangs, but there was no alternative in sight. After many brushes with the police and the law, he eventually landed in prison. His experience there broke him. The harshness of the survival conditions and the exacerbated violence of prison life were not for him. More importantly, his incarceration meant that he had to abandon his toddler son, Carlos. As he was still experiencing the trauma of losing his family during the war, he could not bear to inflict the same fate onto his son.

He attempted suicide, was unsuccessful, and vowed as he was released from prison that he would never "abandon" his son again.

Geovanni's promise to himself and his son came at the same time that the activities of Spanish Catholic priest Antoño Rodriguez Lopez, known by the community as Father Toño, took off in the San Francisco parish of Mejicanos. Father Toño's pragmatic outlook on the daily realities of Mejicanos has endeared him to his community. Over the years, he has never promised them a heavenly afterlife to follow their earthly passage made of hardship and violence. In his sermons, he criticizes the Salvadorian government for not addressing their living conditions, for not bringing them the basic human security that anyone deserves. He denounces the collusion of his own leadership, the Catholic Church, with the Salvadorian government, the remoteness of the church's hierarchy from people's daily plights, and the strings attached to missionary work that tighten the cycle of dependency of the "helper-helped." He states: "I think that the church has lost the capability to ask questions, as if they already know it all, and want to speak too much about everything; I believe the Church, or us who represent it, also need to ask ourselves some questions." He hasn't promised anything in terms of earthly humility and heavenly reward, yet he has enabled within the population of his parish more than anyone could: a capacity of introspection, autonomy, and self-confidence as a community. Through his organization, Servicio Social Pasionista, his local team has initiated projects that first took care of urgent health issues in the community, and then looked at addressing structural violence through the strengthening of social capital. This could only happen through living with the community, engaging with them on a daily basis, and observing more than claiming to know what they needed. It took Padre Toño's organization more than 10 years to gain its momentum and take its current shape.

When Geovanni met Father Toño, he knew that he had finally found a way out of the MS-13. Father Toño became a mentor to him, and with his assistance, he founded a bakery whose primary objective was to provide an alternative activity for other gang members. After an initial financial struggle, Geovanni's bakery became a local success, recruiting more and more youths away from the gangs, so much in fact that the government of the Netherlands has been sending former Dutch gang members to take part in the program and document how this initiative could be replicated back home. While busy with his endeavor, Geovanni founded his own family. His daughter Grace Isamar was born in 2011.

This idyllic picture of hope, social entrepreneurship, and community building was shattered on March 20th, 2012, when he was arrested for having, according to the media, not the police, an alleged connection to murders that had taken place in the community. His bakery was raided, his equipment shattered, and all his colleagues arrested with him. The next day, all were paraded in front of TV cameras as dangerous gang members. Geovanni and his naked tattooed torso made the headlines on TV; he was alleged to be gang-leader "El Crazy" and branded as a dangerous murderer who had been on the run from the police for many months. As a decommissioned Mara, Geovanni was always trying to cover his tattoos from the sight of others, wearing long sleeve shirts and a baseball cap. He was not ashamed of his past, but wanted to look toward the future and wanted others to see him as a human being, not a feared Mara. As I watched Geovanni's public humiliation on TV, the day I was supposed to meet him face-to-face for the first time, I recalled the harsh words of Padre Toño's Sunday sermon a few days earlier: "if God existed, he would care about your misery." On that particular day, Father Toño was not only referring to the God of the Catholic Church's establishment, but also to the municipal and legislative elections that were taking place, asserting that no political party would ever be the solution to the daily issues of the neighborhood. For Father Toño, all political parties make the same promises, only to behave exactly the same way once elected, with utter disdain for human misery. Geovanni had tried to make a difference in his community, yet his life was being destroyed as a pawn to political demagogy. In El Salvador, the invocation of gang violence is what issues such as terrorism or foreign threats can be for Western democracies, a strategic manner to divert citizens away from more pressing domestic issues and inequalities. While gang violence exists, in the same way that terrorism does, it can be conveniently used in populist speeches. With his words questioning God's existence, rather surprising in a Catholic Church Sunday sermon, Father Toño was reminding his congregation that only they were in charge of their daily realities, yet this assertion came with an obvious risk. The day before his sermon, he had received an anonymous phone call: "your life will never be the same again." Communal empowerment, it seems, is seen as a threat to some.

Neither Mara, nor Slave

Awaiting his trial, Geovanni was transferred to the Bartolina prison in Santa Tecla, where, at the time of writing, he is currently sharing a four point five square meter cell with eight other detainees, with only a bucket in lieu of lavatory. Geovanni and his cell mates have to take turns to sleep, are barely fed, and have no privacy. Geovanni has not been granted the right to have any visitor. He has no contact with the outside world and can only hope that the judicial system processes his case as soon as possible to flee his living hell. Letters of support for him have come flooding in from all corners of the world, attesting to his rehabilitation and the respect he has earned within his community and beyond, but these have been ignored by the judicial system.

His legal team has been told that his arrest was connected to a tragedy that shook the Mejicanos community in 2010. On June 20th 2010, a street gang set a bus on fire: carbonizing 11 people inside.[74] The attack was a reprisal against a bus company that refused to give the local gang its monthly protection money. It was the first time that a bus took fire with people inside, and the event shocked the community. The individual arrested in connection to this was proposed a reduced sentence by the judiciary system, in exchange for the names of any gang members living in the neighborhood. Geovanni's name was given, among many others, regardless of his rehabilitation. He was arrested for being a MS-13 gang member, nothing else.

There is a political aspect to Geovanni's "capture", which was planned carefully from a media perspective, with TV crews following every part of his arrest: from the rude awakening he received at the hands of the police, in the middle of the night, to his transport in the back of a pick-up truck.[75] No lurid detail was spared from the public, apart from the police abandoning the screaming three-year-old daughter of one of Geovanni's housemates in the middle the road at 3 am. The Salvadorian government had carefully prepared this media opportunity in the context of its "hard hand" policy against the gangs.[76] Following its election to the Salvadorian presidency in 2009, the FMLN, formerly a guerilla movement against the US-backed right wing Arena government, had to prove to the Salvadorian population that it could be as "tough" on crime as any self-respecting government. This is the political context that led to the arrest and incarceration of Geovanni and his friends. The fact that they were decommissioned did not matter to anyone; they were statistics that could show that even a former guerilla government could be as potent, if not more, as any other. Of course, this policy would also dissuade any member of the Mara from being decommissioned, since it seems that, to the eyes of the world, once a Mara, always a Mara. As for Father Toño, after calling for a dialogue with the Maras as part of a commission that was set up between the Catholic church and the government, he was seen to be taking sides with organized crime, and started receiving anonymous calls and letters.[77]

The election of the FMLN to the presidency of El Salvador was short of being a revolution for many Salvadorians. After years of armed struggle against a right-wing government, the FMLN was finally going to represent "the people". The hope this electoral result generated was immense, matched only by the rhetoric of its charismatic President, Mauricio Funes. Three years on, police repression and judicial miscarriages such as Geovanni's are still taking place on a regular basis. Many feel let down, and have resorted to communal solidarity as their way out of poverty. Geovanni's choice to live his own life, away from the extremes of either being a Mara or a slave for the ruling elite, sweeping streets or cleaning rich people's toilets, was too much for El Salvador's polarized elitist social system to take. The state had to re-appropriate him as a gang member, years after he had chosen a different way of life. Will he be sentenced for living in the excluded middle? How long will he remain behind bars? Despite the admission on part of the police that "El Crazy" is actually another person, the political nature of this case seems to be working against Geovanni, as the government will not want to lose face after it made such a show of his arrest.[78] In the meantime, Father Toño continues to fight for his liberation despite having his daily community activities reduced to a minimum. He maintains that he does not care about being killed, after all, what better footsteps to follow than Monsignor Romero's? Can Geovanni's case be seen as an illustration of El Salvador's aborted social revolution?

On Revolutions

Revolutions do not take place in a vacuum. They are often the product of years of awareness, and at times struggle, toward social change. They conventionally occur when a critical mass pulls itself together to demand change, accountability, and a new social order.[79] Revolutions might occur when enough people are aware, not only of what they want to see change, but also of the fact that this change cannot come to life within their current elite structure. Superficially, revolutions are about numbers. When enough realize that something in their daily reality just does not work anymore, that it does not match their dreams and expectations, and more importantly, that they have nothing left to lose, they take to the streets and demand change.

Change they often obtain. Political figures are deposed; to be replaced by others deemed to be less corrupt, more just, and more representative of the critical mass that led the revolution. Yet the months following revolutions often have a bittersweet aftertaste. They see disenchanted masses going back to their daily grinds, hung-over from the intoxicating flavor of a change that has not challenged their structure of oppression. Leaders change. Cards are re-shuffled. Hopes are confined to legal structures. Elections are organized. Inequalities persist. We are then told that power corrupts. The greater cause that united all slowly withers away toward the replicating of former divisions. Old wounds re-open, and social conflict resumes.

Who can safely state that the 2011 Arab Spring has significantly changed the power structure of today's Egypt?[80] Are women better off than before?[81] Are confessional differences erased?[82] Do ordinary people feel protected by their security services?[83] Do ordinary people have a brighter economic future? Has structural violence diminished?[84] Their elite commentators and media pundits now tell Egyptians that change takes time, that they have to be patient, and that Rome was not built in one day.[85] Do revolutions truly produce social change, or do they merely re-shuffle the cards that always bring the same elites back in power? Are revolutions an end in themselves?

Kuhn writes that scientific revolutions bring personal as well as discipline-wide upheavals that are neither expected nor desired, due to the existence of a paradigm greater than the individual within it.[86] A paradigm is the epistemological foundation of a social structure. It cannot be challenged through logical appeals. Any contradictory element emerging out of that paradigm is thus characterized as an anomaly within the system, an exception to the otherwise uniformly-working order. Thus, when a theory is disproved through an experiment, the scientist's first reaction will be to label it as an anomaly, and often not venture down the new avenues that pursuing it might entail. The reasons for this are the following. First, no one wants the foundations of their scientific existence to be shaken. Kuhn equates the feeling that this creates to one's floor being pulled out from under one's feet. Second, and more importantly, no one wants to be a misfit, or an anomaly, within one's own discipline. In fact, within scientific revolutions, critical mass is often more of an impediment than a force toward creativity and renewal. Kuhn states that scientific revolutions do not occur through the exposure of undeniable scientific facts, but through a leap of faith. Scientific revolutions are not the effect of a logical change, but of a change in emotions. They occur when the scientist cannot but move forward, since the groundings of his/her discipline have vanished; yet a new ground must be landed on. An unavoidable leap into unchartered territory defines a scientific revolution that brings a paradigm shift, a true leap of faith.

Paradigm shifts do not come about cumulatively, steadily, reasonably. They are resisted tooth and nail. Nyhan and Reifler demonstrate that facts do not change one's opinions.[87] Their research suggests that the more politically aware someone is, the more difficult it is for them to accept facts that contradict their beliefs. Essentially, Nyhan and Reifler found that facts have the tendency to backfire and actually reinforce people's political opinions. Should a supporter of the Iraqi invasion, for instance, be confronted with the fact that no weapons of mass destruction were found in post-Saddam Iraq, this politically astute person will further justify the occupation of Iraq.

Critical mass and facts are thus not the prerequisites for a paradigm shift to occur within any approach to social change for peace. Revolutions by numbers - mainly relying on critical mass - have the tendency to backfire and bring more of the same, while paradigm shifts need both a space for creativity and a leap of faith to flourish. Old paradigms can be deconstructed, yet social change can only emerge out of a paradigm shift. Revolutions, such as the Arab Spring, will never amount to sustainable renewal and social change if all their players are not ready to throw themselves together into the unknown.

Easier said than done? A revolution that counts on numbers and progression will only find structural resistance to it, since in essence, elites may give in a little in times of upheaval, but will never relinquish their space. Can paradigm shifts emerge from resistance, struggle, even violence? Is it possible to bring about change outside the system? If the system is too great a force, why fight it? Why not just be the change we want to see around us? Easier said than done indeed. Geovanni tried to change his daily reality, his own being, only to be politically recuperated by the state eventually. For the state, which only sees a gang member in him, his decommissioned self was an anomaly to its polarized right/wrong reality, it could not be left to live outside the parameters of the state, the political elite, and the Catholic Church: once a "bad" Mara, always a "bad" Mara. The threats to Father Toño's life come from this very dynamic.

The proposed solutions to a crisis are always those that we know, socially learned, and part of the same pattern of failures, discontent, and social inequalities that we periodically denounce. We are told to be patient, since democracy will save us in the end. We know that the map is not the territory, but is democracy even a relevant map? Democracy in Ancient Greece amounted to the power of the Greek adult male elite.[88] Women were confined to their homes, slaves were subjugated and barbarians, i.e. foreigners, were not even considered second-class citizens. Is this the democracy that we want to reproduce in Egypt today, that we want to maintain in El Salvador? Is the democracy that is producing Occupy Wall Street protests all over the world the one that represents us, or a select politico-financial elite? Why do we look up to a two thousand year old metaphor to organize our daily lives, while social creativity might have local answers to our present needs? How much more patience must we show? Why not give our collective creativity a space to re-organize our spaces?

By re-organization, do we mean throwing away our old paradigm? Again, in the spirit of Rumi's field beyond right and wrong, old paradigms are not "wrong," they are incomplete, limited, and non-adapted to current realities.[89] The fact that we can no longer ignore the anomalies that exist before our eyes forces us to reconsider our current paradigm, to renew ourselves, and to indulge in a leap of faith. What have we got to lose? An emphasis, not on prayer and hope, but on daily actions, changed daily realities, and anomalies, is the paradigm shift.

On the Immanence of Peace

How does the older paradigm within which we live manifest itself on a daily level? How are we conditioned into waiting for a better life while the structure of our society cannot permit fundamental changes to occur? In the same way we look up to a two thousand year old metaphor to shape our daily lives, we also look up to powerful figures to enlighten us and show us the way to our own salvation. It is in this context that the, for some, shocking words of Father Toño can be best understood. Statues of noble men, and few women, adorn our collective spaces. These men are deemed to be "visionaries", showing us the way toward a better future, toward peace, equality, or freedom. They are strong and assertive; they can also be violent and merciless. They are revered. The anniversaries of their deaths are celebrated, while the peaceful daily actions of many faceless women and men are kept invisible. Similarly, as we remember wars and their ending dates for the sake of never forgetting how low we as humanity can sink, the interludes of peace that we collectively live and work through are not part of our history books. We are thus conditioned into believing ourselves to be inherently violent beings that need great men to wage our wars and negotiate our peace.

When great men are a product of our system, their allegiance to it reinforces our shared illusion that revolutions bring change within the established order. Division within a system is what reinforces society. Hierarchical relationships between those great men and us mere mortals, between us and a nature, us and animals, can all be found within our common social order and come to define our daily lives. None of us are powerless, yet the belief that we are keeps us dominated. According to Boff and Hathaway, the belief in our own powerlessness, our collective analysis, is maintained through dynamics of internalized oppression, denial, addiction, and despair.[90] Additionally, the systemic reinforcements of those states permeate through militarism, our educational systems, mass media, and of course, organized religion. All those sectors of our lives rely on the same hierarchy of human beings that maintain the supremacy of those great men we are told to look up to. From military generals to Hollywood celebrities, through our higher education administrators, and Jesus Christ, the transcendence of the "visionaries" over commoners, of mind over matter, keeps us from realizing our own potential.

To the question as to why these "visionaries" are mostly men, paternalism reassures women into believing that one of them was always the immanent force behind this or that great man out there. First there was the Virgin Mary, and now there is Michelle Obama. Women are also told that their time will come, while being shown that if they were one of the boys, they would earn their own statue within the garden of transcendent visionaries. Does a statue, or even an inversion of domination contribute to social renewal? Patience within domination, again, is exhorted within the system. Does patience make social peace, or does it merely reinforce a status quo? According to the current ratio of seats gained by women in the US Congress, which has not changed since 1979, parity will not be achieved for another 500 years.[91] Is there no space between the status quo and revolution? Is the status quo the only "peace" that both women and men can afford to live by within a paradigm of fragmentation and domination?

Sometimes, the status of great men does not emanate from the ruling elite. A stone in powerful shoes when they are alive, they are eventually granted the halo of "visionaries", preferably after they are dead and buried, as with Reverend Luther King. As they challenged the system outside the polarity of revolution and status quo, they evolved within an Aristotelian excluded middle, the field beyond right and wrong, violent change and peaceful resignation, which can both be seen as part of the same coin.[92] Those "visionaries" living through the excluded middle were referred to as anomalies within the system. Their deaths allow them to be recuperated by it. What can be learned from their lives? Can their lives be re-appropriated by the mere mortals, the foot soldiers, the proletarians, the downtrodden? Can vision and peace be immanent, or are they condemned to the realm of the transcendent? The immanence of social peace is the core of Padre Toño's message to his parish: "If there was a God, he would care about your misery." This assertion does not belong to the realm of theology, but to that of social reality. If the Catholic Church no longer has answers, yet does not dare to ask questions, who will?

Can anomalies play a larger role than we think in order for the paradigm shift to take place? Should the leap of faith that any paradigm shift needs to emerge start from within us? Anomalies may not be an exception after all; could it be that we all have the potential to become anomalies that can gain faith in ourselves, instead of the statues that we are told to revere? Is the crippling effect that the statues have on us, as a collective, be the meaning of the idolatry that is forbidden in some religions? Has the meaning of idolatry been devoid of its core by the transcendence of organized religion? When anomalies are no longer the exception, and when they are defining an emerging paradigm, do we no longer see them as "anomalies" through the overcoming of our own internal as well as structural powerlessness? Does this awareness preclude any social action toward peace?

The recognition of anomalies within one's mind is the beginning of a paradigm shift. What are anomalies within the mind? They arise with our awareness of our injustices to ourselves, the contradictions of our daily lives, and the disconnect that exists between our thoughts and our hearts, and the depression that arises from it. The search for happiness, or more accurately, the end of suffering, may arise from the overcoming of one's own powerlessness, or the faith into our own internal anomalies, highlighting our disconnections between mind and matter.

Physicists Targ and Hurtak argue that suffering derives from one's attachment to a representation of the ego as the self, through the construction of an Aristotelian way of thinking, based on duality: either, or.[93] If I do not enjoy school, I am not intelligent. If I am not with them, I am against them. I have everything to be happy, yet I want more, I buy more, and I desire more. I believe in gender equality, yet I will keep exploiting my "maid" at home. Equality for the chosen few transcends the immanence of equity. According to Targ and Hurtak, awareness of this disconnect might lead to an end to suffering, to inner peace. Inner peace, in turn, might contribute to social peace. Yet, through the either-or logic, I remain divided, disconnected from my fellow human beings and myself. I feel superior to my fellow human beings, through judgment, hierarchies, yet I want to be more in touch with my heart and myself. Beware though, a distorted sense of connection to the self can also be part of the "either, or" paradigm. If I comply with military service in my country, if I occupy others, shoot them, torture them, am part of a state's repression against another people, and yet go to India in search of myself as soon as my military service is over, I am looking for a fragmented inner peace, I am still part of the either-or paradigm: either military, or hippie. Once I have found myself in a great ashram, meditated enough to forget about my active participation of state (structural) violence, I will return home and be a pillar of my society again. My conscience will be silenced, and I will function as a polarized agent of structural violence again.

Structural inequalities, therefore, feed into internal powerlessness, and vice versa, while an emphasis on collectiveness to renew the system remains a question of numbers and not an awareness of, and a faith in, internal anomalies. After all, those who feel the most powerless individually have the most to loose, while the structural maintenance of social domination places an emphasis on individuality. There are as many realities as there are beings on this planet; this realization leads to greater connectivity, structural peace. There is neither right, nor not-right, in Rumi's field beyond right and wrong.

Can we simply say that there is neither peace, nor not-peace? If peace does not have to be transcendent, then can its imminence be nurtured or harvested both at the individual and the collective level? What does peace look like in the "field"?

What Happens in the "Field" and Whose Peace Are We Working for?

The relevant illustration of the peace that can be witnessed in the "field" is, again, Korzybski's "the map is not the territory".[94] The map, in terms of the peace that we are trained to work toward, is all the good intentions that are put together in terms of policies, international law instruments, Millenium Development Goals, and the theoretical models and measures of peacebuilding, such as capacity building. The territory is the reality of what happens on the ground: the shortcomings that arise, the deaths that occur as a result of what was supposed to be a useful policy, the disappointments that appear, and the lack of sustainability of most initiatives. Why is there such a disconnect between the map and the territory? It might be because both arose within a dualistic mode of thinking, that is to say, the absolute belief in the transcendence of peace.

Gandhi's vision that there is no road to peace, that peace is the road, may link the map with the territory, yet, Gandhi's vision is not as important as the statues that have been erected in his name. Through the transcendence of peace, Gandhi remains another visionary consigned to a convenient space in our peace gardens. Idolatry strikes again.

If you are not at peace, you are at war, and vice versa. Within a transcendent vision of peace, meaning that peace is the responsibility of states, the UN, visionaries, and peace practitioners, ordinary people can rest on their laurels and have a Sunday picnic next to Gandhi's statue. It is a transcendent vision of peace that, for instance, categorizes countries that are at war, over those that are at peace. Intervention, good deeds and intentions, will only take place in states that are deemed to be at war, whether internally or with other states. After their nice picnic, ordinary human beings who are committed to peace will donate funds to those poor countries out there that are crippled by war, malnutrition, and despotic leaders. They may shed a tear over the plight of the less fortunate with a moral sense of compassion, and a confortable condescendence. The visibility of war becomes the greatest comfort for those living in peace, just as one feels the coziness of a home while living through a snowstorm.

Yet, the map is not the territory. While most might believe that Africa and the Middle East are the most violent regions in the world, it is actually Central America that qualifies as such, with a regional average of 29 violent deaths per 100 000 inhabitants on a yearly basis, the highest in the world.[95] According the Geneva Declaration on Armed Violence and Development, there has been more violent deaths in El Salvador than in post-Saddam Iraq, with a rate of 60 deaths for 100 000 inhabitants.[96] These figures would confirm that most of the countries that are touched by violent deaths are not engaged in war. Are they at peace? They certainly are considered to be so on a map. The country has been officially at peace since 1992, the year that the Salvadorian civil war ended and the United states government massively expulsed Salvadorian refugees.[97] As a result, the Salvadorian gangs of Los Angeles, the US-based sections of the Mara Salvatruchas, returned to create more havoc in their former home country, resulting in the aforementioned grim statistics. Can peace be deadly and violent too? The transcendence of the Chapultepec Peace Accords in El Salvador certainly has been so for many civilians over the last twenty years.

Peace and war are the two ends of an Aristotelian dualistic model. This model, or paradigm, has been the basis of our modern thinking since Descartes separated mind and matter, and equaled human society to a machine.[98] If you are not with us, you are against us, and if you are not at war, you are at peace. Nothing exists in between. There is a strong relationship between cause and effect when it comes to "peace" from a map perspective, what William Easterly refers to as the "planners' peace", and Oliver Richmond as "liberal peace".[99] According to this mode of thinking, peace comes with a few tools and recipes that will ensure its success. Most peace and conflict trainings worldwide teach these tools to an army of soldiers either coming from peaceful/democratic states, or elites of the developing world that have a vested interest in maintaining a social status-quo that benefits both them and Western democracies. Before looking deeper into this assertion, let us examine not only some of the results of mechanical peace, but also how its dualistic structure maintains conflict in the long run.

The case of Kashmir lends itself to the analysis of peace as a map, or a machine, as well as the anomalies that this paradigm creates and the conflicts that it maintains. First of all, the internationally visibility of the Kashmir issue has always been in relation to the two nation-states that are said to be in dispute over its territory, namely India and Pakistan.[100] Seldom in the international media is the voice of the Kashmiri people heard, for if it was, the dualistic nation-state frame covering it would be broken over the possibility of a third option: an independent Kashmir. Is this third option a panacea though? Or is an independent Kashmir the recipe for yet another conflict down the line? The situation of Kosovo today, struggling to exist as a nation-state with strong ethnic minorities, amidst regional tensions and a global economic crisis, is not an encouraging precedent. Second, two of the main solutions conventionally proposed to solve the Kashmiri issue are also deeply entrenched within the Cartesian frame of the nation-state. Those are plebiscitary and partition approaches, respectively leaning toward an allegiance to India or Pakistan, or creating an international border between India and Pakistan, alongside a specific political community of Kashmiris.[101] Seeing Kashmir from the vantage point of either-or carries a strong dualistic pattern that has led the conflict to be deemed intractable for years. Both solutions are deemed by Sumantra Bose to be obsolete, since they were based on a geopolitical paradigm that no longer exists; yet it is alongside those roadmaps for peace that the United Nations presence in the region has been based since 1949, through the UN Military Observer Group (UNMOGIP).[102]

After more than sixty years, one can safely ask: for whose peace has the UNMOGIP been working? Equally dubious, under these dualistic nation-state parameters has been the stance of the Kashmiri local political elite, siding alongside India or Pakistan, or merely calling for independence. Can everyone find themselves in the elite that represents them? Or is it the responsibility of everyone to find a solution to his or her daily problems? Can peace in Kashmir only take the form of the nation-state? Or do the peace roadmaps brought forward for so many years merely maintain a status quo that keeps killing ordinary human beings on a daily basis?

The death toll of the Kashmiri conflict, on the Indian side, is estimated to have been between 50,000 and 70,000 since 1947.[103] The ratio of security forces to inhabitants of Kashmir is one to six, hardly the hallmark of what is believed to be a democratic state. Communication in Kashmir has been restricted since the Delhi central government banned SMS messages a few years ago, roadblocks restrict people's movements, demonstrations are often crushed with live ammunition, and individuals can be arrested on suspicion of involvement in anti-government activities as established by a Kashmiri version of the post September 11th 2001 Patriot Act in the United states, the Army Special Forces Act.[104] The violence of Indian state repression is not confined to Kashmir, but is broadly applied to other nationalist movements, peasants, so-called "dalits" or untouchables, and various ethnic groups, all in the name of upholding its sovereignty as a democratic nation-state.[105] In effect, India, as most other democratic states, puts forward a homogenous front that is jealously guarded in the name of its sovereignty, thus creating layers of visibility and invisibility in its projection of itself as a democracy.[106] Anomalies again take the "invisible" form of arbitrary arrests of "opponents", which can be ethnic groups, political factions, or minorities, often leading to their torture, brutal repression, and/or arbitrary detention. However, since those anomalies are deemed to be invisible, the democratic civil peace theory put forward by Gleditsch dares to contend that, in addition to not fighting each other, democratic states maintains civil peace by upholding the rights of each and everyone of its citizens.[107] Go tell that to Geovanni and Father Toño.

Democracy pertains to the transcendence of peace, while in effect, the model that it represents crushes the invisible, represses its anomalies. This becomes a reinforcement of the dualistic model of liberal peace through democracy. In this sense, democracy and the civil peace theory is an offset of a mechanistic paradigm that attempts, with utmost failure, to explain a static reality that is all but homogeneous, since it is highly complex, shifting, and as diverse as the human race. When the civil peace system is given a higher value than people, between 50,000 and

70,000 people die in Kashmir under the umbrella of democracy and a UN observer mission.

If democracy fails to cater for a complex non-dualistic reality, how about peace missions? Fisher and Zimina have sought to ask this very controversial question a few years ago, in an article called "Just wasting our time?"[108] It is not the first occasion this question has been asked in the peace and conflict studies circle, yet this article was widely circulated within the field, as it systematically reviews anomalies both within and resulting from peace missions, from the strings attached to the state funding of NGOs and INGOs, to the unhealthy competition between peace projects, and the questionable training and motivations of peace workers. Overall, this article claims that mission statements often remain within the dualistic vision of negative and positive peace, contained in turn within an overall transcendent peace-building vision, meaning that peace may only be established through the promotion of certain values and projects emanating from the global North. Whose peace are those missions working for: their funders, the survival of their projects-based funding, their own administrations, their value-systems? When Fisher and Zimina contend that most vision statements rarely address structural violence or work toward deep cultural change, their narrative touches debates over revolutions and social change in general, reshuffling cards but not encouraging a paradigm shift. While questioning the industry's dismissal of conflict as inherently bad, they challenge a dualistic vision of peace and conflict, arguing that conflict may also be a force for constructive transformation. They make the case for transcending a technical approach to peace, within which peacebuilding is understood as fixing a car, again, challenging the Cartesian paradigm of peacebuilding. Yet, what happens in the field is mostly the expression of a highly fragmented peace, whose visible deliverables hide a forest of anomalies within which unsustainability reigns. Is this state of affairs the result of a cynical approach to peace, or the mere reflection of a dominant epistemology of control? Are the good intentions of peace workers sufficient to build peace, in this sense? Or is the notion that peace can be built fundamentally flawed?

If the leap of faith toward paradigm shifts and the immanence of peace may be a reference, then the idea of "building peace" maybe the greatest flaw of the peace industry. Fisher and Zimina advocate a transformative approach, Richmond a post-liberal one, Lederach an elicitive method, and this book, a decolonized epistemology.[109] Much remains to be assessed within a spirit of complementarity, yet decolonizing peace requires a leap of faith. Could a closer look at the complexity of anomalies, from a non-mechanistic perspective, prepare the way for such a revelation?

Conclusion

Dysfunctional categories of peace not only illustrate domination, they maintain it, through the narrative of transcendent peacebuilding. Those dysfunctional categories are based on emphasizing the duality between good and bad, the people worthy of being helped, and the others, falling outside the realm of assistance because they are not "worthy" victims, because they do not know their place, or because they want too much too soon. A linear thinking of cause to effect prevails in the dysfunctionality of peace building, whereby the prevalence of an outside interference is revered. Peace is presented as a sum of isolated elements in a neutral space, and equals to planting a tree in a concrete base. Benchmarks and impositions measure the window of time within which this tree can grow, and they do not take into account the environment within which they are being imposed. When violence epitomizes the state Paradigm, which maintains social peace, how can it not also prevail in peacebuilding, and why are we even surprised when peace operations turn violent toward the populations they are supposed to protect? The next part of this book will look into the complexity of anomalies from a non-mechanistic perspective, and will attempt to enlarge the scope of peace away from it being a mere reflection of the mechanistic paradigm of domination and control. A different pair of lenses will be used to understand why some projects, concepts, ideas that may be called hybrid by some, actually are sustainable in their nature. Deconstruction uncovers anomalies, and their analysis enables us to comprehend the paradigm shift that has already taken place. It is time for us, peace practitioners, to join in.

III. Anomalies and the Invisible

In January 2001, four months into my PhD, I decided to go to Lebanon for my "field research". I was supposed to spend at least a year of desk research beforehand so that the reality that I would find on the ground would be strongly situated within the existing parameters of Lebanese Studies. I had read all that I could find about Lebanon in my Irish university, and was determined to find out what was really going on in the "field." Nothing could have prepared me to encounter such a different reality. As soon as I arrived in Beirut, I could not recognize or match the world I had read about with my surroundings. I set off to find a way to meet the Hezbollah with apprehension, hoping I would be able to meet them after a few months of intense networking on the ground. After all, they had been described in all the literature that I had read as a highly secretive organization. To my surprise, I found their details in the Yellow Pages, and was soon given a *carte blanche* to become acquainted with any of their social or media initiatives for the next two years.[110] A fascinating experience came a few weeks later, when I visited the Irish army in their camp in Tebnin, South Lebanon. As part of their outreach initiatives toward the population of South Lebanon, they had decided to broker a farming initiative between Irish NGO Bothar, USAID-funded Mercy Corps International, and Hezbollah-funded Jihad al-Binaa. The project was supposed to enhance farming capacities in the little town of Barachiit, with the innovative twist of linking all parts of the population, Christian and Muslims alike, in the pursuit of a common goal.[111] I was so enthusiastic about it that I shared this with some journalists as soon as returned to Ireland, only to be scolded by the Irish Army spokesperson for publicly linking the Irish Army to the Hezbollah. No matter what was happening down South, back home, I was not expected to run against preconceived notions of what the rest of the world was supposed to look like. The Hezbollah could not be seen as anything else than a group of religious fanatics. I was given a stern warning, should

I write about any of this, I could kiss any future collaboration with the Irish army goodbye. Somehow I did not mind, and in the years that followed, I have never looked back.

The emergence of anomalies in a socio-political context can be as destructive as it can be productive. While anomalies represent the expression of deep issues within societies, they can be met by a variety of reactions, ranging from blatant attempts to suppress them to a resigned acceptance. When those anomalies are understood to be challenging the mere existence of the paradigm within which they emerge, they often tend to be subdued. Names are quick to be invoked, for those anomalies to be conveniently labeled and returned to their rightful place in the deviant spheres of the paradigm, the unmentionable invisible. Life then returns to its normalcy, until the anomaly re-emerges, stronger than before. How can these anomalies be understood in their complexity as well as in relation to what they are revealing to us about the system within which they emerge? Should anomalies be understood in an alternative way? Can they shed light on what makes them alive, sustainable, complex, and unique? Once they are understood, can they be integrated within society's core? Can they make a significant contribution to society? Revolution or reform, is that the question? In this chapter, I will look into three different anomalies, to understand what they have, or have not brought to their surroundings, what elements of sustainability they harbor, and how their complexity can contribute to a wider debate on decolonizing peace.

The Lebanese Hezbollah, the Anomaly's Visible

The Hezbollah as we know it today was not "created" as such, rather, it came to exist as an "organizationally amorphous and rudimentary constellation of primarily Shi'i Islamic activists cells dedicated to fighting the Israeli army of occupation."[112] Those cells were composed of all sorts of individuals and groups, some following religious leadership, others more secular, some fighting a small-scale insurrection against the Israeli presence in South Lebanon, others involved in political resistance to social disparities in Lebanon, etc.[113] The growing discontent within part of the Lebanese Shi'i population was expressed in the elaboration of various groups, which, in the early summer of 1982, gelled together.[114] This was a sensitive period of time when Lebanon was occupied by Israel, the Lebanese government was thought to be representing only the interests of the Christian ruling elite, and the traditional Shi'ite leadership of AMAL was increasingly resented for allegedly siding politically with the country's elite.[115] Shi'a Muslims in Lebanon had traditionally been at the bottom of the social scale, until they lived through a collective revival in the 1960s, symbolized by the charisma of Imam Musa Sadr, who traveled extensively through South Lebanon in his Volkswagen Beetle to spiritually, politically, and socially empower its Shi'a population.[116] Imam Sadr soon founded the Movement of the Disinherited, but disappeared in Libya in 1978. The movement's armed wing and subsequently gentrified political outlet, AMAL, saw many of its more humble members become disenfranchised without the charisma of their popular leader. They chose to leave the party and form their own groups, spiritually aligned with the 1979 Islamic revolution in Iran and Lebanese Shi'a clerics.[117] Those groups, over the years, gathered into the formation that published its first manifesto in 1985, under the name of Hezbollah, the Party of God.[118] Still, some associated groups came to prominence in October 1983 with suicide attacks against the American Marine compound and the French paratrooper barracks in Beirut, killing 299 soldiers, six civilians, and two suicide bombers. Even though the presence of the French and US as part of the Multi-National Force (MNF) was labeled as a peacekeeping operation, both France and the US soon became drawn into the Lebanese civil war, openly siding with the Lebanese Christian political elite against other parties to the conflict.[119] Not only did the MNF use its naval power against Druze fighters above Beirut, it also launched air raids against training camps near Baalbeck in the Bekaa valey, where Iran was deemed to be training the nascent Hezbollah.[120] While an organization called Islamic Jihad claimed responsibility for the attack, denouncing the French and US involvement in the conflict, it is widely believed to have been part of the future Hezbollah.[121]

It is in this context that the Hezbollah has never ceased to be labeled in the West as a terrorist group, embodying most clichés that pertain to such type of organization. The public image of the Hezbollah is seemingly inseparable from indirect actions that took place more than thirty years ago. By way of a comparison, one can imagine the state of Israel constantly being referred to in the mainstream media as a rogue state, due to the bombing of the King David Hotel on July 22nd 1946, killing 91 people and injuring 46.[122] Mass media reports and academic books, often named "Inside" the Hezbollah, as if it was difficult to consult the Yellow Pages, rarely relinquish the sensationalist temptation to introduce the organization with the mention of the attack on the US Marines, followed by a string of well-known polarizing truisms including training camps, secretive hierarchies, charismatic leadership, Islamic fundamentalism, martyrdom, undignified mothers ready to see their children blown to pieces, anti-Semitism, chanting fanatics, the state within the state, etc.[123] The mainstream of media reporting on and academic analyses of the organization are in fact situated within a frame that makes any complex understanding of the organization virtually impossible. Why indeed would anyone seek to understand the Hezbollah beyond the string of clichés that will make one's media program popular, or one's research project attractive to state funders, where the money actually is? As I mentioned earlier, the emergence of anomalies in a socio-political context can be as destructive as it can be productive. In the case of the Hezbollah in the 1980s, it was destructive, yet its persistence, flourishing, and gentrification over the years can also allow us to understand what makes an anomaly sustainable, and perhaps more importantly, how its integration within society can allow it as a whole to thrive. Anomalies never emerge and sustain themselves out of a social vacuum; an understanding of their invisible aspects may therefore inform us of what transformations a decolonizing peace and conflict studies may be able to undergo.

Complex Adaptive Systems

What conceptual techniques could be used by a social scientist seeking to understand an organization outside its polarized/dual context? Foucault writes on the situation of knowledge from an epistemological perspective.[124] What we see is informed by the conceptual techniques we are using. Thus, the evolution of knowledge within a Cartesian framework will inevitably taint any analysis according to the polarized parameters of good and bad, from a mechanical perspective. My contention is that an analysis of the Hezbollah with a mechanical, hierarchical outlook might be extensive and undoubtedly scholarly, but will invariably fail to bring to light its most important features, the invisible ones. It might mention those features, yet its dualistic/mechanistic lens will never allow them to be seen outside of a Cartesian context. [125] Alagha has previously established some steady grounds on which to be built by describing the Hezbollah as a "sophisticated, complex, multifaceted, multilayered organization."[126] My own contribution to his pioneering work is to argue that the Hezbollah is more than the sum of its parts; it has formed and maintained itself as a complex, adaptive system.

Educator Donald Gilstrap writes about different types of organizations that range from Cartesian/rigid to more organic/adaptive systems.[127] According to this vision, equilibrium-oriented systems are the more rigid type of organizations. They work as a closed circle, with limiting negative feedback loops that are solely bound to prevent change and move toward the stability of their point attractor, which bounds the "system to a stable position of rest."[128] The negative feedback loops dampen and regulate "activity to keep it within a certain range," and can be compared to a thermostat.[129]

Those organizations typically consume more human energy than they generate, since they operate in a closed circle. That would be a typical hierarchical and rigid structure that may attract new membership but will invariably fail to sustain itself in the long term due to its rigidity. The very mention of "sustainable peace" as a contemporary issue highlights the failure of many peace initiatives to generate their own energy, to recognize that what make peace is people, not budget lines, structures, or hierarchies. A typical expression of an equilibrium-oriented system would be a project, NGO, or an organization that looses its momentum once its funding runs out; it can also be a project whose administration has taken over its core *raison d'être*, message, etc. The UN would be a prime example of an equilibrium-oriented system, while its blue helmets' sexual abuse of local populations would be the symptomatic expression of the burial of its moral compass under the weight of its ever-growing administration, its point attractor, hence the unsustainability of most of its "field" programs and initiatives.[130]

An alternative to this is the complex adaptive system, which, as its name suggests, describes an organic structure that is in constant adaptation and symbiosis with its environment. It manages a simultaneous state of order and disorder, while it exchanges energy with the outside world. It is a system with an ability to continuously re-emerge at more complex levels of development after having reached certain bifurcation points, which enable it to transform itself. Flexibility is one of the most important elements of this type of organization; it combines the presence of positive and negative feedback loops interacting with one another, and their patterns of connection are expressed in the emergence of a strange attractor. Instead of being geared toward a point attractor, which will keep everything in its place and be predictable, complex adaptive systems revolve around an attractor that is referred to as "strange" since it is a combination of "parameters [that] provide a boundary from which the system does not stray, yet the object's movement within those parameters cannot be predicted within the framework of time or space."[131] Since the system is open to its environment, the behavior patterns of its feedback interactions cannot be predicted – hence the "strange" nature of the attractor – yet its patterns of behavior do appear to follow a certain flow.[132] To be able to grow and sustain themselves, organizations of this type depend on a combination of negative feedback loops, that will maintain their structure, and positive feedback loops, that will push the system into renewed courses.[133]

The Formation and Nurturing of a Strange Attractor

Taking an example from the private sector, a strange attractor can be formed and maintained through the nurturing of three elements: a shared vision that provides an alternative to traditional strategic planning, team processes that allow for the emergence of attractors, and constant, transparent information flows that act as positive feedback mechanisms. To expand on the idea of a shared vision as an alternative to strategic planning, the elaboration of different future scenarios around a common vision are thought to allow each member of the system to "interpret" the different possibilities envisaged within it from the point of view of their own knowledge and experiences. No future course is set in stone, with rigid steps laid out to attain it, rather, space is left for unpredictable parameters to enter the future equation without necessarily derailing a "strategic plan," since what matters is the common vision. This vision allows the organizational dynamic to continue flowing regardless of external disruptions. Team processes, another element that is vital to the emergence of a strange attractor in this context, are characterized by their propensity to auto-organize, share their vision and responsibilities, and auto-regulate. The experience of Dee Hock, founder of the Visa Company, a modest horizontal structure that bases its entire activity on logistically facilitating trust between banks worldwide, is an illustration of how team processes are a vital element of complex adaptive systems. Indeed, team processes guided by a shared vision and an aversion to hierarchies have made Visa one of the most renowned financial institutions worldwide.[134] Finally, constant and transparent information flows are not only associated with the emergence of auto-organization, they are also paramount in keeping motivation, a sense of belonging, and a collective vision intact.[135]

In terms of fostering the emergence of a complex adaptive system, chaos theory refers to three non-linear phases.[136] The least complex of these is often described using the metaphor of turbulence, where uncertainties and doubts are able to flow into a collective space, generally as a result of the rigidities and dissatisfactions inherent to a point attractor system. This phase emulates a degree of freedom after the letting-go of a cripplingly rigid structure. Examples of this may be the chaos of a revolution, or the avid reading of a writer before entering a more creative phase. Once uncertainties and chaos have been allowed to enter the common sphere, and have been structured by a combination of negative feedback loops with specific social, spiritual, or internal parameters, there is a phase of bifurcation and amplification where the turbulence will form a vortex whose combination of feedback loops will gather around a set of parameters that provide both a boundary and space for unpredictable movements, symbolized in the strange attractor. Then comes another phase, characterized by an open flow of creativity and a nurturing of the complex adaptive system. This process can be seen as non-linear and non-hierarchical, as the complex adaptive system is in constant interaction with outside influences. Thus, this depicted cycle of evolution remains in constant flow, and should the internal adaptability of the system become rigid, then it will move toward an equilibrium-oriented shadow of itself, or a hybrid system symbolized by the emergence of a periodic attractor: which may allow for cosmetic flexibility but not for structural change. A democratic revolution, as understood in chapter two as a reshuffling of political cards, illustrates this type of in-between system.

Hezbollah's Invisible: The wilāyat al-faqīh

While the visible face of the Hezbollah as an anomaly can be seen in most journalistic and academic depictions of the organization, its invisible nature is far more complex than anticipated by most self-appointed specialists. What if the Hezbollah was a complex adaptive system? Revisiting the concept of the formation of the Hezbollah not as a self-proclaimed hierarchy of individuals but as a gradual aggregation of communities over the summer of 1982, one understands better how the organization emerged from a network of individuals to a community of like-minded groups with one shared vision: social and political equality for the Shi'i of Lebanon. Originally, the Hezbollah existed primarily as a military group with the immediate goal of ridding Lebanon of the presence of what it considered to be foreign invaders: the US, France, and of course the non-recognized state of Israel, routinely referred to as the Zionist entity.[137] Its long-term goal to "[give] all [their] people the opportunity to determine their fate and to choose with full freedom the system of government they want" has been less salient both in the media and academic literature.[138] This objective referred to the Lebanese consociational system of democracy, which was based on sectarianism and not on proportional representation, favoring the Christian Maronite population over the Shi'i, the largest group.[139] Understandably, the title of the first Hezbollah manifesto, published in 1985, refers to its community as the "downtrodden," rising against social, political and economic oppression, and calling on other parts of Lebanese society, namely the Christian Maronite working classes, to join their ranks as downtrodden. It is in these terms that the Hezbollah can be understood as an anomaly, bursting out of a perceived unjust, locked, point attractor system that can no longer be sustained. This phase in the history of the Hezbollah can be referred to as a turbulence phase, coming as a result of the point attractor of the Lebanese political life, its elitist consociational system, which at the time was symbolized for some Shi'i by the perceived secularist gentrification of AMAL against the best interest of Lebanon's Shi'a population.[140]

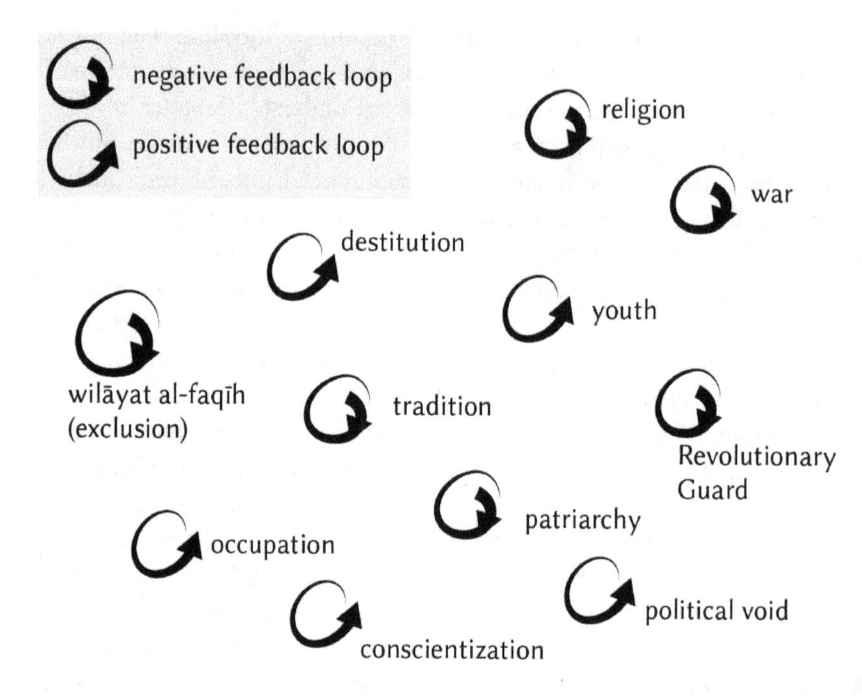

Figure III-1: Hezbollah as a turbulence

Within this turbulence, an aggregation of negative and positive feedback loops slowly emerged (see Figure III-1). Of primary importance to the Lebanese Hezbollah from before it declared itself to the world in 1982 to its first bifurcation point was the concept of the *wilāyat al-faqīh,* namely the 'guardianship of the jurisprudent.' This concept emerged in 1978 from Iran in the writings of Imam Khomeini, and was seen at the time as a revolutionary concept linking the political and religious spheres for the first time in the history of Shi'i Islam.[141] While much international attention was placed on the relationship between Iran and the Hezbollah, in terms of military training, financing, and political support, a detailed understanding of the *wilāyat al-faqīh* has seldom entered the international academic sphere, let alone the "inside

Hezbollah"-type media reporting, yet it is at the core of any complex understanding of the party.[142]

This concept places the political within the spiritual, rendering the highest religious authority the title of "Supreme Leader." According to this model, Islamic jurists, whose role before was confined to "settling legal disputes, collecting religious taxes, implementing statutory penalties (…) and administering the finances of minors and the intellectually impaired" became "invested with much wider powers, including (…) the undertaking of jihad, overseeing the process of individual rectitude and social conformity, and, (…) responsibility for government," all under the overarching guidance of the Supreme Leader.[143] Of importance to this concept are not only governance aspects, but also the spiritual features of individual integrity, social conformism, and humility, in emulation of the humble and pious lifestyle of Supreme Leader Imam Khomeini. Within a Lebanese political context of clientelism, nepotism, and corruption, the *wilāyat al-faqīh* as a spiritual-political concept was nothing short of revolutionary, a rupture with the political tradition and system from which the Hezbollah as an anomaly emerged.[144] This epistemological break, alongside a strong element of armed resistance, was still characterized within a relatively closed structure whose parameters were fixed by the Iranian concept of *wilāyat al-faqīh*. As the 1985 manifesto stipulated, the Lebanese political structure as such was seen as illegitimate, hence while the Hezbollah was granted a significant amount of popular support, its religious element was far from an overall liberal Lebanese social tradition. A strict understanding of religion, in fact, may have transformed the turbulence into a rigid near-equilibrium, whose point attractor was a zealous understanding of religion. The negative feedback loops were close to asphyxiating the Hezbollah as potentially organic system. Many authors picked on the bigotry that existed then, manifested in the closing of coffee shops, prohibition of alcohol, the harassment of women in the streets of Beirut, etc. As Jaber recalls: "[w]omen who were considered to be dressed in an improper manner were often harassed by the radical newcomers – Hezbollah militiamen – and rumors circulated that acid had been thrown at girls dressed in an 'un-Islamic' way as a lesson to others."[145]

Hezbollah's Transformation As a Complex Adaptive System

In the early 1990s, the party found itself torn between a strict abiding to Iran and its wilāyat al-faqīh, or an existence as a Lebanese entity, not abandoning the *wilāyat al-faqīh,* but interpreting it within some contextual parameters. These were times of political renewal in Lebanon, following the signing of the 1989 Taef peace agreement, which marked the end of civil war. While all militias were disarmed as part of this agreement, the Hezbollah was spared as the Israeli occupation of South Lebanon made it a resistance movement in the eyes of some, despite the contestations of others.[146] As a result of these accords, it was decided within the group that its identity ought to be legitimized by as many people as possible within Lebanon, not only their community. This led to the Hezbollah's participation in the 1992 parliamentary elections, since the consociational system requires any political party to forge alliances with other sects within the country. Alagha characterizes the Hezbollah's relationship with the *wilāyat al-faqīh* according to three phases.

In the first two phases, between 1982 and 1992, the group considered the elaboration of an Islamic state in Lebanon to be their primary goal, modified somewhat in the second phase, which focused on how to deal with Lebanese affairs. The third phase saw the party change its interpretation of the *wilāyat al-faqīh*, with the blessing of Imam Khamenei, Imam Khomeini's successor, in order to participate in Lebanon's political life, signifying a shift from "exclusion to inclusion," from a desire for Islamic revolution to one of reform within the Lebanese political system.[147] This shift, coupled with the expansion of the party's social programs, the opening of their TV channel al-Manar, and other efforts to reach out to the larger Shi'i and Lebanese population, allowed for the turbulence to reach a bifurcation point that enhanced the complexity of the aggregation of loops and allowed for the formation of a complex vortex-like structure (see Figure III-2).

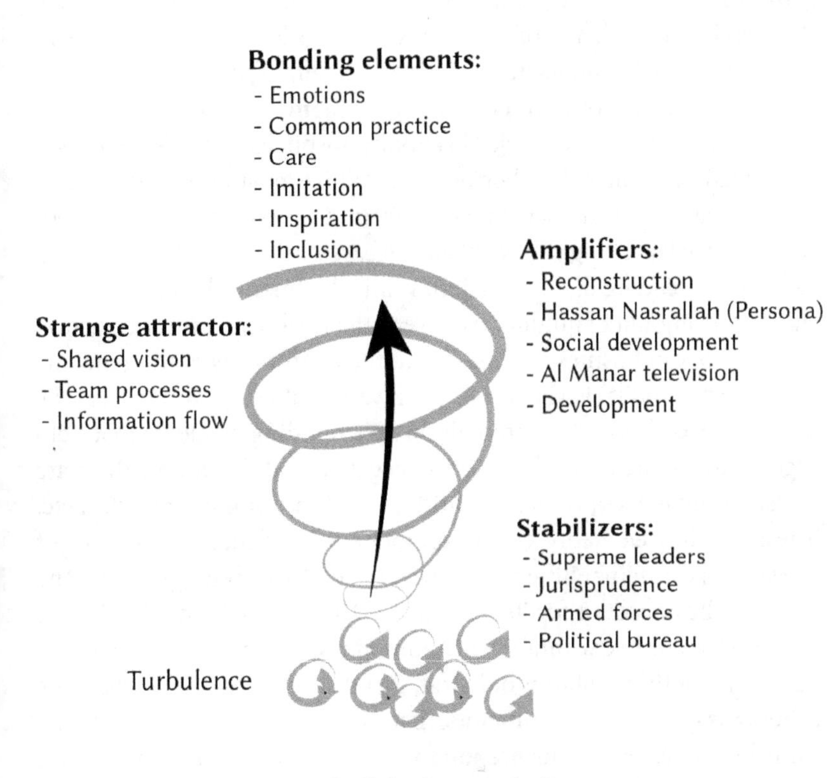

Figure III-2: Hezbollah, from turbulence to vortex

The *wilāyat al-faqīh*, whose originally strict interpretation made it a negative feedback loop among others, evolved into a strange attractor, whose spiritual parameters allowed for the Hezbollah as a system to be in constant movement, developing an internal flexibility in relation to its incorporation into Lebanese political life. Its adaptability allowed for energy to flow into the system, and also for it to renew itself. The Hezbollah as a complex adaptive system was in full movement.

Nurturing the Hezbollah's strange attractor was a shared vision of political reform, a team process within which the promotion and development of its civil society elements was paramount, and a constant flow of information through Friday sermons, its media apparatus, social gatherings, and specific groups such as the Women's Section, etc. While the stabilizing elements of the vortex can be seen as its clerical basis, its armed branch, and its jurisprudence, its amplifying elements are its political inclusion of both card-carrying members and sympathizers across sects, its vast network of schools, hospitals, rural development organizations, cultural gatherings, etc. While most academic debates on the finances of the party revolve around the material help of both Iran and Syria, as might have been the case in the past, this is now an incomplete reading of the situation. While the Imam Khamenei sends a significant amount of funding every year, the Hezbollah owns a considerable number of businesses throughout the country, generating income for the party, and it also benefits from investment ventures, as well as the Islamic tax of *khum*.[148] Within all this are bonding elements that hold together the structure of positive and negative feedback loops. These are invisible elements of care, inspiration, emotion, common practice, etc. These bonding elements, which cannot be quantified, are at the core of the system's sustainability; they are the generators of energy within the system. They account for its endurance. The fact that most Hezbollah political leaders live within their communities, not in separate mansions but in the southern suburbs of Beirut, that their children attend the same schools as those of basic militants, that they all receive treatment in the same hospitals, brings inspiration to the least successful economically, and maintains the bonds between party members.[149] The fact that the party's charitable organizations look after the least fortunate in providing them jobs is also a source of invaluable support. The party is perceived to be connected to the population it represents, and this is what matters. Looking at the Hezbollah as a complex adaptive system enables us to understand it as an anomaly that has been able to sustain itself for years. What if its chaotic elements, such as its strange attractor, its loops, and its bonding aspects, could be made available to the formation of peace as an inherently sustainable system? What more, in the evolution of such a system, can we learn about its sustainability?

The Anomaly's Sustainability As a Complex Adaptive System: A Given?

A complex adaptive system is by definition alive, organic. Has the Hezbollah ever emerged as an even more complex organization, as could be expected by Gilstrap's third phase, characterized by the open flow of creativity? In 2009, the Hezbollah published a new manifesto describing a "momentous time full of change" and the transformations that the party has made while facing it.[150] It clearly reaffirms its willingness to become part of the Lebanese political system and calls for it to be reformed into a consensual democracy, in order to proportionally represent the Shi'i community of Lebanon, the overwhelming majority in terms of numbers.[151] While the evolution of the Hezbollah over time is undeniable, if one studies it beyond the terrorism frame commonly used both in the media and academia, elements of auto-organization and the supposed lack of hierarchy can certainly be called into question. While the presidency of the party, through the position of Secretary General, was initially thought to be rotating as a way for the party avoid being weakened should its leader be killed, Sayeed Nasrallah has occupied this position since 1992, more than 20 years. Has this consolidation of power hampered the party's complex adaptability? Since Imam Khamenei named Sayeed Nasrallah as his religious deputy in Lebanon in 1995, one might conclude that his role is now perceived to have gained a spiritual edge, within the realm of the *wilāyat al-faqīh*.[152] In any case, apart from the Secretary General, positions continue to rotate within the spheres of party, where one never remains in one post for an extensive period of time.[153] This is thought to limit the potential for nepotism. In terms of auto-organization, it is difficult to ascertain, from an outside perspective, if those dynamics are pertinent to the daily functioning of each part of the system.

Has the inclusion of the Hezbollah within the Lebanese political scene changed its popular appeal to ordinary Lebanese? Has the party become less dynamic in recent years? As a primary opposition party within the

Lebanese political system, the Hezbollah failed to convince a significant part of the Lebanese population in the June 2009 legislative elections. Those elections, following 537 days of political deadlock within the government, did not bring the Hezbollah political alliance a majority of seats in the Parliament. A few months later, the party issued its new political manifesto as another strong statement of its commitment to Lebanese political life. Its strange attractor was nurtured in terms of information flows, but will this be enough to reach another bifurcation point? Will this be enough for the Hezbollah to convince a larger part of the Lebanese population of its inclusion within Lebanon at large? In the meantime, how will the system face up to the challenges that could render it more rigid, a hybrid point-attractor system, or worse, a near-equilibrium system? When the invisible elements of the anomaly are made obsolete, its sustainability evaporates.

Letters from Abbottabad

In the eyes of the Hezbollah, no book chapter ought to join discussions on them and al-Qaeda, since both organizations are highly different in their mission, their political philosophies, and, more importantly, their spirituality: the Hezbollah pertains to Shi'i Islam, and al-Qaeda to Sunni Islam. Both branches of Islam are now undertaking a low-intensity conflict within the Muslim world. None of the political events occurring in Afghanistan, Bharein, Iraq, or Syria can be understood without taking this religious war into account.[154] Given this context, it would be trivial to see any collusion or collaboration existing between the Hezbollah and al-Qaeda. The two organizations have much more to divide them than bring them together, yet they can both be understood as anomalies. In terms of their organizational structures, they ought to be compared in order to contribute to this chapter's discussion of elements of sustainability pertaining to complex adaptive systems.

In 2006, I was asked to write a chapter on al-Qaeda for a book on non-violence and counter-terrorism. I accepted the challenge, but thought it would be impossible to reconcile counter-terrorism and non-violence in the same publication. That was until I realized that acts of terrorism are perpetrated by insurgencies, and that al-Qaeda considered itself at the time as a global insurgency. I concluded my chapter with a quotation of British Home Secretary John Reid, who warned that al-Qaeda might be winning the war of ideas in the Muslim world.[155] At that time, many signs showed the potential sustainability of al-Qaeda since it had become an idea, rather than a traditionally hierarchical dyadic system. As its ideas spread across the globe through the speeches of Usama bin Laden, its cells auto-organized into planning and carrying out their own operations, and popular support seemed to be gaining momentum. In this sense, al-Qaeda could be understood as a kind of turbulence made of both negative and positive feedback loops (see Figure III-3).

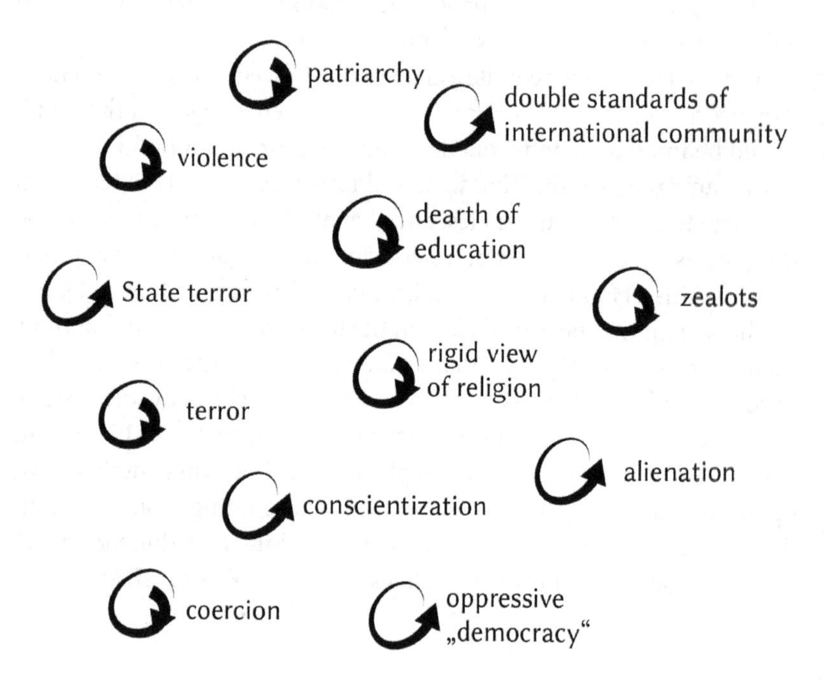

Figure III-3: al-Qaeda's vortex until the mid 2000's

What happened in the period between 2006 and 2012, during which Usama bin Laden was killed, and the organization failed to sustain itself? Months before his death, Usama bin Laden wrote a series of letters in which he recognized mistakes made by some parts of the system that precipitated a massive loss of popular support, especially the indiscriminate killing of other Muslims, the strategic mistakes, and the needless terrorizing of the general public.[156] Are those mistakes, and the killing of Usama bin Laden, enough to understand the demise of the system, and the failure of the turbulence to reach a bifurcation point that would have formed its strange attractor, and made it into a vortex?

At the onset of the Islamic Revolution in Iran, French philosopher Michel Foucault wrote a series of reports in which he presented a strong analysis of the "political spiritualism" of the Iranian Revolution, anticipating its sustainability due to the alchemy between metaphysical and nationalistic dimensions, fusioned in the *wilāyat al-faqīh.*[157] His analysis was dismissed at the time as "infantile leftism," yet he succeeded in showing the relevance of understanding the sustainable features of the system, and by default the Lebanese Hezbollah, from a chaos theory perspective.[158] Another one of Foucault's analyses posited an "analogy between democracy and Shī'ism, on the one hand, and Sunnīsm and tyranny, on the other."[159] While I am tempted to agree, as such an analysis would account for the loss of public support to al-Qaeda between 2006 and 2012, I would do so in more subtle terms, referring to the space for flexibility within the complex adaptive system of the Hezbollah, as opposed to that of al-Qaeda, the presence of many positive feedback loops with the Lebanese social structure, as well as the strong establishment and nurturing of bonding elements within that system.

While al-Qaeda had undoubtedly formed a turbulence in 2006, its decline was due to the fact that its Sunni Muslim precepts were interpreted in a rigid manner by its supporters, who spent disproportionate amounts of time enforcing unpopular measures such as the trimming of one's beard, the banning of jeans, deemed too westernized, the killing of female goats whose genital parts were not covered, or, its all time low, the banning of the purchase of suggestively-shaped vegetables by women, such as cucumbers.[160]

Moreover, there was no bounding with local populations, only fear of summary executions at the whim of uneducated Islamic thugs. In terms of amplifying loops, al-Qaeda in Iraq never brought a strong network of social support. In fact, it seemed to many that they were only bringing fear, death, and destruction, hardly the basis for popular support.[161] While Shi'i Islam is strictly geared toward jurisprudence, which makes it relevant to everyday issues and circumstances, Sunni Islam seems to be limited to the dry precepts of the Qur'an, thus giving space for zealots to ban women from buying cucumbers. Can there be any formation of a strange attractor in a Sunni context? The rule of the Muslim Brotherhood in Egypt will hopefully begin to answer this, and hopefully challenge Foucault's assertion. In relation to our understanding of anomalies and chaos theory, al-Qaeda, at the time of Usama bin Laden's death, had morphed into an equilibrium-oriented system (see Figure III-4). Its leader had understood this, as his last letters indicate.[162]

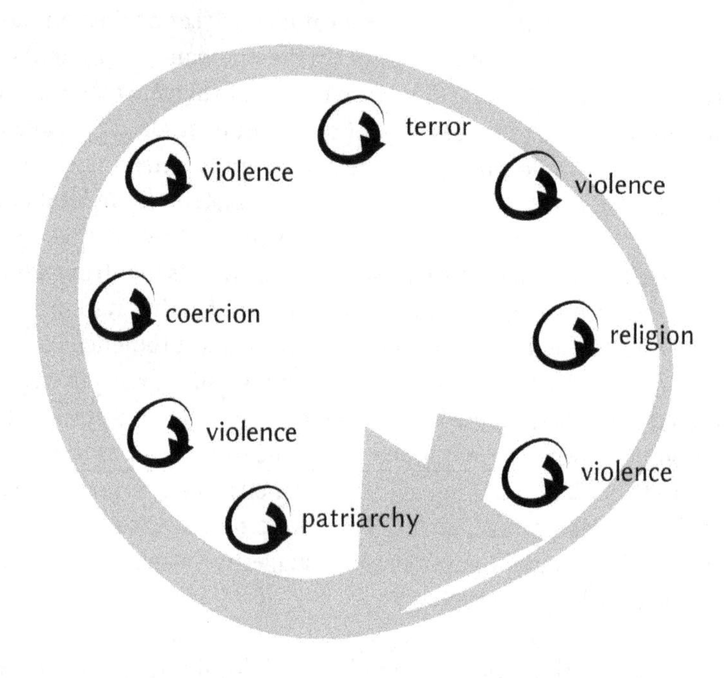

Figure III-4: al-Qaeda as an equilibrium-oriented system

So far, anomalies have been described and analyzed in a violent context; what can anomalies within a non-violent framework bring to our unfolding discussion of decolonizing peace theory? What other invisible parts of anomalies can complement our organic conceptualization of sustainability?

An Ordinary Woman, Almost

Sampat Pal Devi is an ordinary Indian woman.[163] She was born in a remote village of Uttar Pradesh, was married off to a farmer at the age of twelve, and gave birth to the first of her five children when she was fifteen. Nothing in her life was meant to be of any relevance to the outside world. She was born and raised to be a dutiful wife and mother, serve her in-laws, and above all, abide by her social caste and system. From a very early age, Sampat challenged some of the expectations placed on her. She never understood why her brother would be allowed to go to school, and while she would be expected to stay behind helping her parents. So one day, she decided to follow her brother to school and started eavesdropping on his alphabet lessons. She reiterated the experience as often as she could, until her uncle caught her and decided to help enroll her as a fully-fledged pupil. This was Sampat's first act of rebellion and victory, the first of many. Upon reaching puberty, she was deemed eligible to share her husband's household, and moved to her in-laws. Her relationship with her mother-in-law was never one of love or respect. As a new bride, her status was that of a house-slave, and her mother-in-law was violent toward her. As the conflict between them escalated, she rebelled again and again, and was sentenced to death by her new family. This common practice in rural Indian households amounts to banishing the rebellious wife to a garden shed where she will die of starvation. Sampat did not accept this fate, and appealed to village elders to mediate between her and her in-laws. By then, her husband had also taken her side against his parents. All this was unheard of. As a daughter-in-law, she was supposed to be graceful in starvation, but then again, Sampat was "special." The mediation awarded Sampat's husband his share of inheritance from his parents, and they left for another part of the village to start a new life.

Since she knew how to sow, read and write, she started a school where she was able to share all her skills with all the women that sought education. She was giving back the chance that her uncle had given her all those years ago. As a young couple exposed to rural life outside the security of a family sphere, life was not easy for Sampat and her husband. Nevertheless, all started to fall back into place after a few years, almost. Her exposure to the world outside of a family structure, coupled with her traumatic experience of domestic abuse, made Sampat more aware of what was occurring around her, to her neighbors and loved ones. She realized that her plight as a young bride in a violent household was part of the structural violence prevalent within the Indian caste system. Her fate was that of millions of others. She saw Brahmin, people of higher castes, spoliate members of lower castes; she witnessed humiliations, public beatings, etc. One day, she happened to pass by when a Brahman was hitting a man of lower caste, a Chamar. She asked him to stop, he didn't, and before she could even think about it, she hit him. Realizing how far she had gone, she fled to the Chamars' quarters and rallied more than a hundred women to follow her down to the police station. On paper, India is a human rights heaven, all fundamental rights are respected, the caste system is banned, and it is illegal for anyone to hit someone from a lower caste. In practice, the caste system is still rampant and violence prevails. After reaching the police station with her group, Sampat asked to register a complaint denouncing the act of violence she had witnessed earlier. After refusing to oblige, the policeman was shown the group of supportive women sitting outside, and finally agreed to lodge the complaint. This was a victory, not only for Sampat, but also for the lower castes of her village, which collectively lost their fear of the higher castes and even elected a Chamar as village representative, an unprecedented move.

Once again, Sampat did the unthinkable: she ate at the new village representative's house. According to the caste system, since she had consumed the food and water of an untouchable, she too had become one. Following a series of other similar moves, in which she assumed she would make people realize the absurdity of their traditions, it was considered that she had gone too far. Both her in-laws and the higher castes of the village decided to put a contract on her head. They contacted five *dada,* hit men, to kill her. Sampat was unaware of this until one of them, a distant cousin, greeted her and asked about the sewing teacher in the village. She replied that she knew who this woman was, and he then confided that he was to murder her with four of his colleagues, for the sum of 10,000 Rupees, about $180. She decided to invite the five of them for dinner before revealing that she was the person they were looking for. After a lavish meal, she disclosed her identity, asking them to listen to her full story before deciding to kill her or not. As she explained her past actions, which challenged the structural violence that they too had been subjected to during their lives, they unanimously decided to decline the contract that had been offered to them. The same night, Sampat decided to leave town and start a new life elsewhere, once again. She realized that this was a close call, and that she might not be so fortunate the next time.

The Great NGO Scheme

Her new life was to be set in the little town of Badaosa, close to her parents' village, where she first opened her learning center, and later her own self-help women's group. This self-help idea was *en vogue* at the time, as was the founding of an NGO. Sampat's group was not established to carry out international or government-funded projects, and had no fixed monetary compensation to offer anyone, including her. Rather, it informed women of their rights in terms of inheritance, as well as domestic and civic life. It brought women together in solidarity with one another, and it also navigated the Indian bureaucracy around land reform laws, passed decades before, ensuring that their application would be to the benefit of the lowest castes. Sampat, in a way, was more of a social worker or a lawyer than classic NGO worker. One day, as she was helping a group of women to open bank accounts, she met a man called Jay Prakash. He was working for his own NGO, gathering groups of women opening bank accounts on the basis that with their joint funds, they would later be able to secure loans for investment in the local economy. This was supposed to be a win-win for everyone involved, NGOs, banks, and the government, and of course women, would be able to expand their businesses, while paying a minimum of 20% interest rates. Jay offered Sampat a job with his organization and she accepted. For every group she would put together, Sampat would receive 500 Rupees, and for every loan secured with a bank, Jay's NGO would receive 2000 Rupees.

Sampat was a natural at gathering women under this government scheme, as they had complete trust in her. In a few weeks, the NGO had become eligible to secure more than 100,000 rupees. This was never to materialize, however, as Jay's two other workers had registered many of the groups with a separate NGO that they had established themselves. Since Jay had put all his savings into his NGO to keep it afloat until the government money came, he was left with nothing. The few women groups that both he and Sampat managed to keep to their name never obtained the government reward, as their case government officer wanted a cut in their earnings. A sad realization dawned on both Jay and Sampat: their women's groups would be indebted for years as a result of this scheme; it was not a win-win after all. The financial autonomy encouraged in Sampat's original scheme was replaced with the burden of debt. Sampat helped Jay back on his feet with one very clear lesson learned: they would never open a traditional NGO and would never focus their community activities on money. What this episode had taught them was that none of the people actually supposed to benefit from these schemes ever did. There was no individual or community empowerment in this.

Gulabi

A new beginning was called for, in the same town, with the same women, but this time on the basis of all but monetary values. Sampat's extensive network would be completed by Jay's astute experience of the Indian social and legal system. Joining efforts, both would be able to have a deeper impact in their community. As Jay put it: "I knew she was capable of greater things than I could ever imagine, so I stepped aside and decided to support her in whatever she would undertake." Many cases came to Sampat and Jay as news spread by word of mouth, and their network soon grew beyond the frontiers of their district. The cases that came to them were varied: an abusive husband, a corrupt police officer, a rogue NGO scheme, etc. One day, she was called to a nearby village to help after a local Brahman had decided to build his new house not only on communal grounds, but also, and more importantly, on the only pathway that joined one end of the village to another. One foundation wall was already built: Sampat decided to destroy it. She realized that her strength was her network of men and women, and that due to cultural issues, women would never be arrested *en masse* for destroying the wall, which they did in a few hours, before going to the police the next day and denouncing the building as illegal.

After this episode, Sampat realized that the group not only needed a name or recognition: it needed a symbol. She chose a pink Sari, the only color that was not political, and the *latee*, the traditional weapon of her caste, a wooden stick. From then on, the group would be called the Gulabi Gang, as *gulabi* means pink in Hindi. This was a bifurcation point; as soon as women would gather, their colored uniforms and name gave them a sense of collective recognition. Would others take them seriously? The case that gave Sampat and her network recognition in their community was when they were asked by two of their members to tackle the issue of government subsidized food being illegally sold by corrupt government stores employees. In India, the poorest are issued a Below Poverty Line card by the government, which allows them to buy subsidized food at a lower price than that of the market. Often, there is no grain in government stores, having been resold at the markets. What Sampat and her gang did was to follow the trucks that were moving the food, highjack them if needed, and hold them until the police arrested the culprits. Many were arrested in connection to this, and the police and district magistrates eventually grew impatient at all the extra work. One day, the police arrested two men connected to a government store, at the result of a routing fight, but kept one of them detained for ten days because he was the husband of a Gulabi Gang member. Sampat and her women came to the local police station demanding the release of this man, and when thugs came to intimidate her and started to hit her, she used her *latee* against them. As a policeman came out to arrest her for the violence she used against the thugs, she hit him with her *latee* too. Soon, the thugs and the policeman were surrounded by dozens of armed angry women, in plain sight of other police officers who never came to help their colleague. Sampat and her gang had again done the unthinkable; another bifurcation point had been reached. Powerful men were scared. Even though Sampat was going to have trouble with the law, all now knew that she and her group were capable of resorting to violence in extreme cases.

An Open Circle of Help

Since this famous incident, the Gulabi Gang has been evolving quickly, with more than 60,000 active members in the network, one-third men and two-thirds women. All members have to pay 100 rupees to become part of the network, a substantial amount of money for people of modest resources. With this amount of money, women receive their pink sari, and the help that they need from Sampat and her gang. This is the only condition for getting the help they need, which according to Sampat "is not a one way street." She explains: "many NGOs give, give and give, and everyone is used to taking without ever offering anything in return. This does not help people on the long-term, keeps them in a cycle of dependency." From her perspective, only in helping another does one realize that he or she can challenge the basis of structural violence within society, since with this help comes the consideration that one will never be alone in "doing the right thing." Not every member of the gang needs help from Sampat. When she is not busy with a case, she organizes rallies in the center of villages to expose people to the gang. To promote her network, she does not use typical tools with pie charts, speeches, or testimonies: she uses local folklore. She sings her own songs with everyone learning the words as she goes along. She organizes plays where the social situations she encounters most are played and remedied on the spot, her own instinctual theatre of the oppressed. Those gatherings can last hours, and their pinnacle remains the self-defense training that she gives, teaching women how to use their *latee* against violent husbands, potential rapists, etc.

One woman was not so fortunate as to have received that training. I met Lila Wati when she was seven months pregnant. She had come to Sampat's house with her ten-year-old boy. They were beggars. Seven months before this, as she had recently become a widow, she was raped by two Brahmin while working in a field. She became pregnant, and her husband's family used this as a pretext to shun her and her son. She was thrown in the streets, and first came to see Sampat when she was five months pregnant. Sampat offered to help her press charges against her aggressors, but all she wanted was money to survive with her son. Sampat gave her 1000 rupees, told her to find a place to live and to come back soon. She never returned until the day we met. She was in a desperate state this time, and asked for more money from Sampat. As I listened to their conversation, I felt an overwhelming urge to help, to make things better for her and her son, to take her in my arms and tell her it was all going to be fine. I discreetly asked Sampat if I could give her money, and she became furious. After two days with her, I still had not understood anything. She said to both of us: "how dare you ask for more money, and you, do you think that throwing cash at her problem will help her? She might want help, but she does not want to help herself. " I realized I was completely off the mark. Sampat offered for Lila to stay with her until her baby was born, to help her press charges against her aggressors and to train her as a seamstress so that she would be able to make a living for herself after her baby was born. As a counterpart, she asked Lila to be a gang member and to help others when the need arose. Lila refused, she replied that she preferred to be a beggar at religious festivals, that she was making good money, and that she enjoyed her freedom. A short-term monetary gain against a long-term ability to sustain herself: it was her choice entirely. Sampat told her that she was not ready to help herself through helping others, but that her door would always be opened to her. This was the greatest lesson that I could ever learn. Assistance has to come at the right moment.

A Self-Organized Chaos

More than just an interesting story, the episodes of Sampat's life illustrate a cycle of renewal that eventually leads to the formation of a vortex. The selected details shared above show a series of bifurcation points in the life of Sampat that have added to the complexity of the turbulence around her. An aggregation of positive feedback loops provoked a persistent thread of change and renewal in her life – first learning how to read and write, then moving out of her in-laws, then moving out of town, then her failed NGO venture – and these, combined with the positive feedback loops of her constantly challenging the frontiers of her condition and that of those around her, added complexity to the turbulence (Figure III-5), and resulted in the formation of a vortex once she had met Jay Prakash, identified her network as her primary strength, and realized that an open circle of help was the sustainability of her network (Figure III-6).

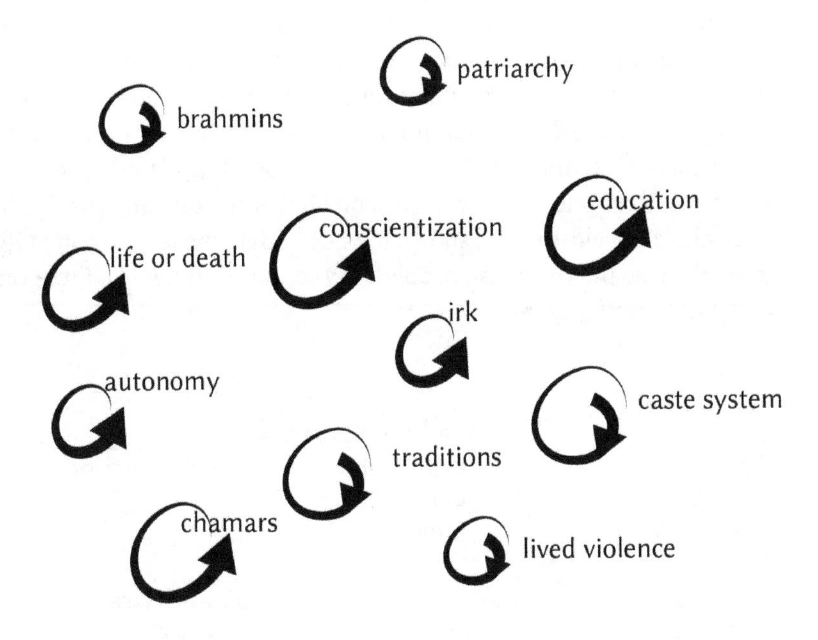

Figure III-5: Sampat's life seen as a turbulence

The coupling of both negative and positive feedback loops resulted in a bifurcation point with the realization that the network's social fabric was more important than any money-driven NGO, and favored the emergence of a strange attractor, nurtured by the circle of help, the self-organized nature of network, and the symbolism of the color pink. Some stabilizing elements to the network include the presence of people such as Jay, who put their long-term experience of India's civil society scene toward the amplifying of the structure, and the fact that the network is indigenous to that particular part of the Indian Uttar Pradesh. Another stabilizing element, interestingly, is the Indian legislation, which legally sanctions all actions of the gang. Nothing that they are doing actually falls outside the remit of Indian law, apart from, maybe, the aforementioned incident with the police officer. That incident established an aura of deterrence to the network, which could be interpreted as an amplifying element, alongside the word to mouth, networking, folkloric meetings, the history of the network, and of course, creativity. Sampat's creativity in playing outside the boundaries of her caste and tradition can also be seen as an amplifying element. In relation to bonding elements, we find advice between members, compassion shown for one another in the open circle of help, the care shown to one another, the love in humanity, mutual trust, and the constant solidarity for one another. All these can count as bonding agents.

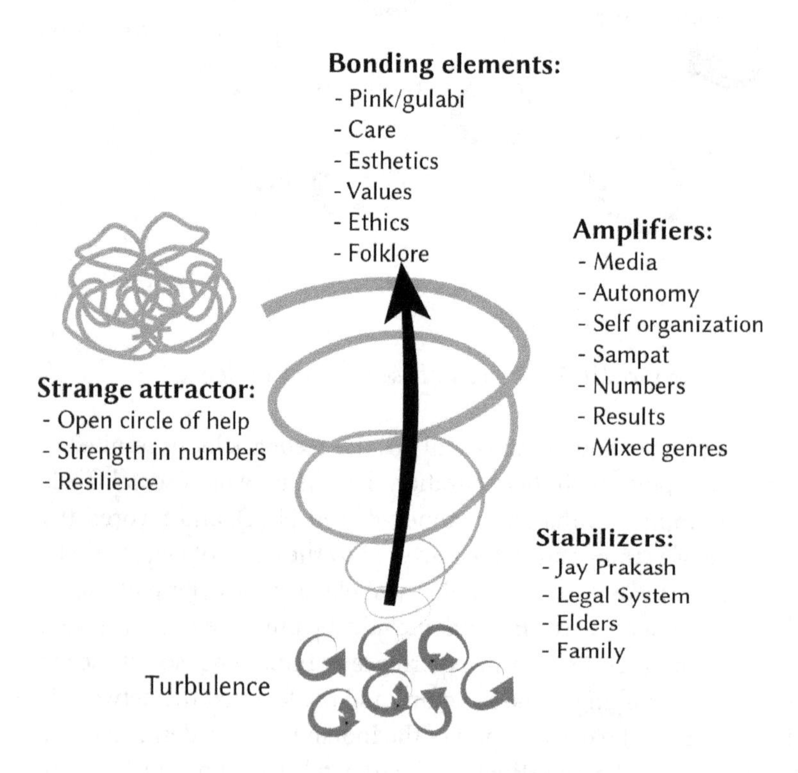

Figure III-6: Gulabi gang as a complex adaptive system

Who is Sampat in the network? Is she a leader? The network, now 60,000 strong, cannot be understood as a traditional hierarchical organization. Sampat does not centralize the actions of the gang, nor does she edict orders to her members. Women in different districts carry out different activities, on the basis of the creativity shown by Sampat in her first public actions. Not only would it be impossible for Sampat to channel everything, but it would also necessitate an important logistical component that the network does not have. While Sampat receives new cases in her office of Attara, other women gather themselves at one another's houses in other parts of the district. Every week, different actions take place simultaneously in different corners of the district, all independently of one another. The one common denominator of the network is equity. From that perspective, the network is self-organized, leaderless, and without hierarchy.

Formative Events and the Panarchy of Living Systems

As narrated above, one may see the story of Sampat over the years as an up and down rhythm of formative events that have made her what she is today. However, this up and down narrative can be seen as a Cartesian perspective of her life, disregarding the invisible elements that have all been as seminal as the most obvious incidents of her life.

As a theoretical addition to our understanding of how anomalies generate and sustain themselves, the panarchy of living systems addresses the sustainability of complex adaptive systems in relation to the interactions they maintain among internal and external factors, as well as the constant patterns of their evolution.[164] Panarchy places the system in relation to its larger environment, and addresses the inner-workings of that system. From this additional perspective, the life of Sampat and that of the Gulabi Gang are inter-related with one another, and embedded within a village, district, regional, and state context. Recognizing the interactions between all these factors gives a deeper understanding of not only what sustainability is, but also how decolonizing peace emerges as a complex adaptive system within a given context. From the perspective of panarchy, Sampat's life is seen within the Gulabi Gang, which is seen within its community, etc. (Figure III-7).

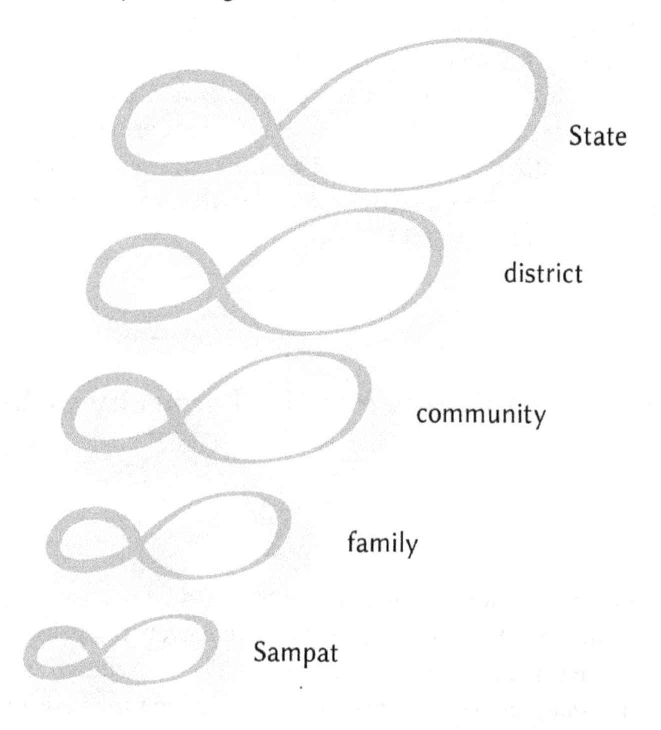

Figure III-7: Sampat's life seen within the panarchy of living systems

A helpful metaphor may be a set of Russian Matryoshka dolls. The hierarchical sense, from this perspective, must not be understood as a top-down chain of control, but as "semi-autonomous levels (…) formed from the interactions among a set of variables that share similar speeds."[165]

The lowest levels of this hierarchy are always the speediest in relation to their ability to change and transform, such as Sampat's life, as opposed to the highest levels, whose more complex structures slow down their ability for change and adaptation, such as the traditional caste system or modern Indian state within which Sampat continues to evolve.

The stabilizing and amplifying elements of a complex adaptive system are respectively seen in panarchy as the internal controllability of a system and the inherent potential of a system that is amenable for change. While the former relies on the "degree of connectedness between internal variables and processes", the latter relies on the "wealth" of the system in terms of the range of future options that are possible for its evolution (Figure III-8).[166]

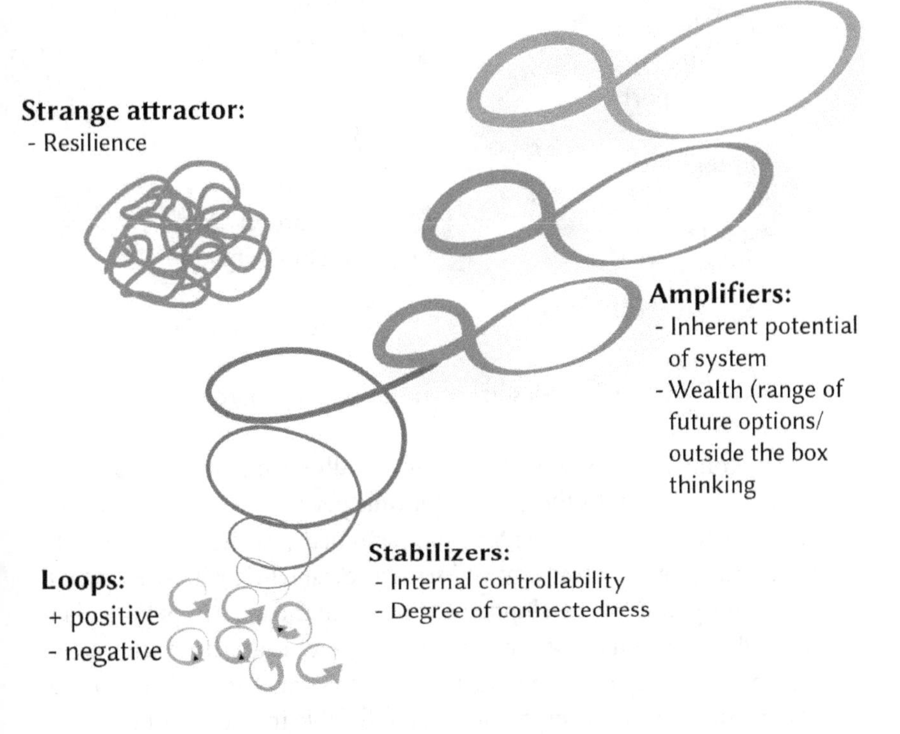

Figure III-8: The morphing of a vortex as a panarchy level

The coupling of both forms a cyclical dynamic of long and predictable periods of systemic consolidation, marked by a slow accumulation of wealth and resources, an increased connectedness, and eventually a rigidity that will be followed by shorter periods of renewal where the potential for opportunities and space for innovation is at its greatest (Figure III-9).

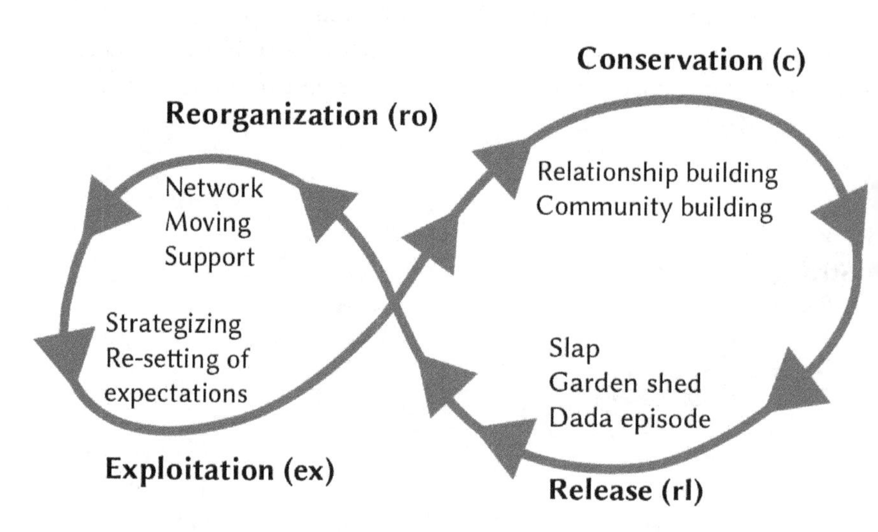

Figure III-9: Sampat's life seen through a panarchy cycle lens

Any complex adaptive system will cyclically encounter those periods. In relation to Sampat's life, this longer time period would be the consolidation of her relationship with her husband in the earliest part of her life, the building of a community and strong social fabric with the *Chamar* community, and the development of ties with the women of her village through her lessons of sewing and writing, etc. That period, characterized by a lengthy time span, also enabled Sampat to realize that her community was allowing her to remain flexible in her personal life. At the culmination of this process comes the phase of rapid reorganization after the larger traditional structure of both her village and community sentenced her to death, involving the unpredictable interference of outside actors, the *dada* episode, which called on the flexibility both of her person and her immediate community network to renew itself elsewhere but within a similar space and time.

From the perspective of panarchy, transformation, change, and renewal are parts of any organic structure's life, which is understood to be in perpetual evolution (Figure III-9).[167] It is, in fact, the resistance to or acceptance of organic evolution that renders the inevitability of change into either a long-term positive or a negative occurrence. Should Sampat have resisted her village elders and remain within their community after the infamous *dada* episode, she would probably have died eventually at the hands of other *dadas.*

Sustainability comes in as the capacity to create, test, and maintain the adaptive capability of the system through what has already been introduced as a strange attractor, which also bears the hallmark of resilience. Within the emergence of a strange attractor, the resilience potential of the system allows it to adapt to unpredictable external shocks. Thus, the phase of rapid reorganization of the system can be seen as one of "creative destruction" where the potential for innovation in preparation for the next cycle is heightened; in other words, the "adaptive cycle opens transient windows of opportunity so that novel assortments can be generated."[168] Looking of the life of Sampat, this particular phase would include settling in to a new part of her village, or thinking with Jay about how to transform a resounding failure into a success by changing some of the epistemological parameters of their relation to one another, as well as to their networks of women, i.e. loosing the money element (Figure III-9). From the perspective of decolonizing peace, any "intervention" outside this specific cyclical window, say, at a slower phase of increased connectedness and rigidity within a particular social situation, would be rendered useless at times, and also potentially counterproductive, hence Richmond's notion of resistance to traditional peace-building.[169] From this perspective, one can assert that the systematic understanding of a post-conflict situation being a universal open window for change is erroneous, as it fails to take into account the cyclical stages of the Matryoshkas.

Revisiting Sustainable Peace

Sustainabilty depends on interactions among internal and external factors, in the same way that all levels exist in constant interaction. There are two stages in which those levels are directly related to and influential toward one another. Looking at the interactions between Sampat, her community, and the local state, one can see how Sampat's constant and more rapid evolution will influence the community when it reaches a level of consolidation, thereby influencing the phase of creative destruction. This is called the "revolt connection". As this phase is enabled, the larger, slower level of the state will influence the reorganization phase of the community through a "remember connection" that will enable the community to re-think itself according to previous trial and error experiences that can account for the memory of the structure (Figure III-10). Take the French Revolution and the phase that followed the revolt connection, joining the people and the social structure, yet failing to connect to any larger level. While the revolution was able to occur as the manifestation of an anomaly within the French political system, the absence of a slower, larger level above it created a constant state of revolution that led to the state of *Terreur*, or terror, which saw the killing of more than 100,000 people in less than two years.[170] From this perspective, the absence of a higher level can lead to a constant state of unrest, failing to add complexity to the adaptive system. Equally, the absence of a connected and more dynamic lower level can lead to a constant state of remembering during the reorganization phase of an intermediate lower structure.

After a while though, a dysfunctional cyclical dynamic that either keeps lower intermediate levels in a state of remembrance, or higher intermediate levels in a state of revolt, will bring one of two types of dysfunction to the system, respectively rigidity traps and poverty traps (Figure III-11). The former will temporarily disable any state of creative destruction or release from occurring due to the stiffness and calcification of the connectedness phase, while the latter will lead to an inability of the structure's social fabric to re-form itself.

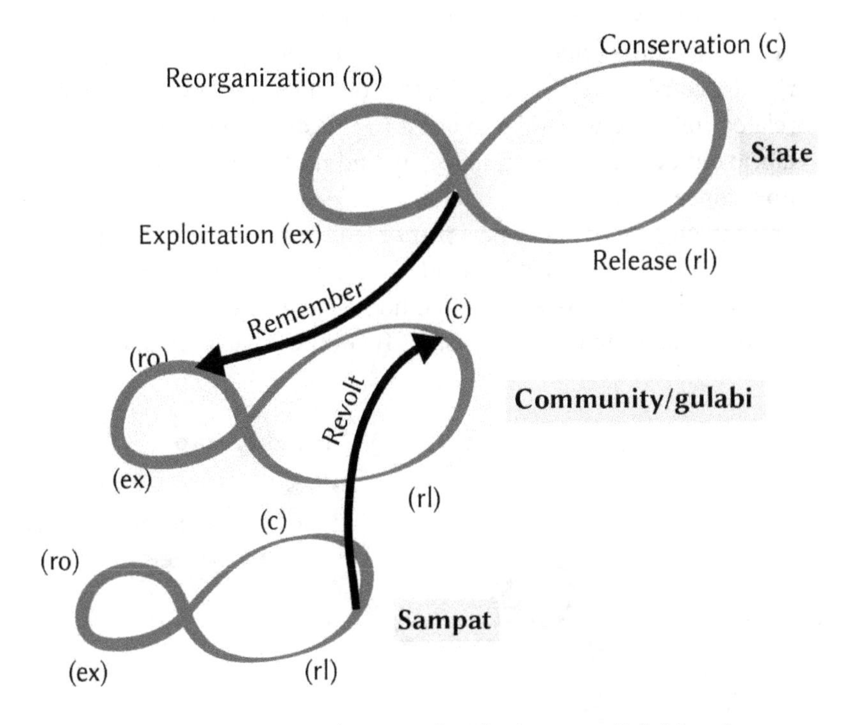

Figure III-10: Interaction between levels: Sampat, Gulabi and state (based on Holling 2001)

An illustration of a rigidity trap can be a police regime in which any opposition that threatens the connectedness of the structure will be met with violent repression, while that of a poverty trap can relate to the dearth of social connections, possibilities for innovation, the existential that will prevent the connection between reorganization and exploitation. An illustration of a poverty trap could be the difficulty of a strong secular civil society in a local-local context to re-form itself after the fall of Saddam Hussein, due to the fact that it had not been allowed to thrive during Saddam's dictatorship. Rigidity traps and poverty traps are not an end, as no cyclical evolution of any complex adaptive system may be halted. While it may take years for a rigidity trap to be broken, since this phase can be longer than others, it will invariably evolve toward a phase of release. The longer it lasts, the harder the fall will be and the more difficult it may be for a poverty trap not to replace its rigidity counterpart. Once again, the role of any peace or development related initiative would be to assist in the strengthening of the social fabric out of the poverty trap it may find itself. From this perspective, no engineering may occur; only assistance to what is already emerging, present, or developing. A new pair of lenses is necessary to logistically accompany a process that is already underway. It is from the perspective of a complex adaptive system that no sustainable peace project can be undertaken: it can only be assisted. That difference is crucial.

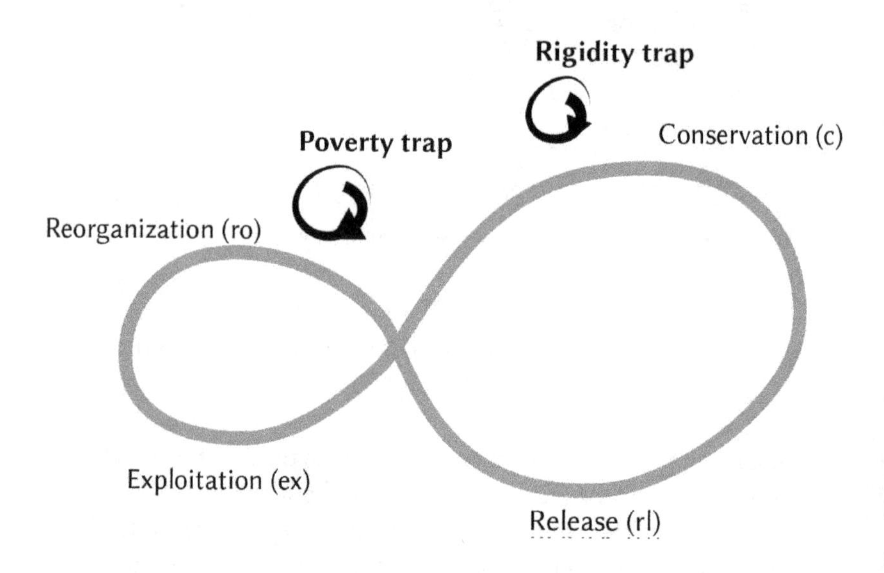

Figure 11: Panarchy and rigidity traps as maladaptive
(based on Holling, 2001)

Including the Anomalies into the Formation of Peace

In relation to peace and conflict studies, the post-liberal peace dialogue has recognized not only the limits of state building as a process, but also liberal peace as a paradigm. Adding substantially to the theory of post-liberal peace, Richmond has coined the idea of peace formation as a space within which local-local actors, such as Sampat, involved in

> *"peacebuilding, conflict resolution, development, or in customary, religious, cultural, social, or local political or local government settings find ways of establishing peace processes and the dynamics local forms of peace, which are also constitutive of state, regional and global hybrids. They may do so in relation to local understandings of politics and institutions, welfare and economics, social and customary resonance of identity, law and security, framed also by external praxes of intervention."*[171]

This notion recognizes the legitimacy of local agency in the "mutual construction of the local, state, and international." How does Sampat's network, and the complex evolution of her life and community, fit the sphere of peace formation? While her strength lies in her constant ability to challenge structural violence, her networking skills and the life of the gang as a complex adaptive system rest on the legal basis of the Indian state and society. Sampat's role, as a local agent of peace formation, is to bridge the map and the territory (Figure III-12). The state/legal level allows for the anomaly that her self and her gang represent to influence a system of which she is seeking to be a legitimate part. From this perspective, decolonizing peace is enabling local-local complexities to inform a larger hybridity.

From a panarchy perspective, peace formation is a "remembering" higher level that keeps decolonizing peace from running in "revolting" circles. Decolonizing peace is therefore a lower level of the peace panarchy, able to break out from cycles of exploitation and conservation faster than higher levels. It exists and can only maintain itself within the existence of higher levels based on similar values.[172]

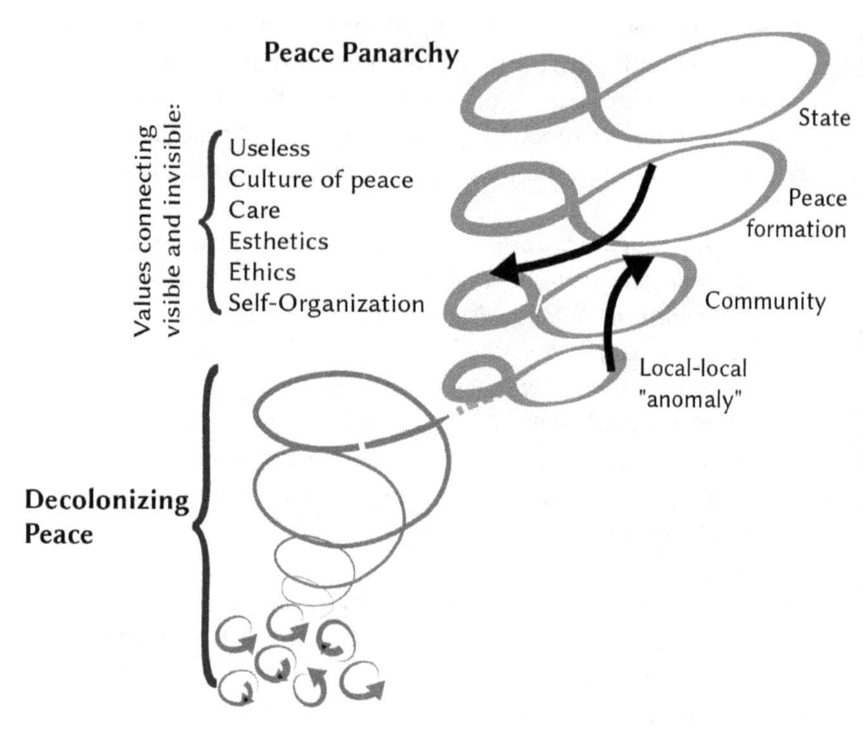

Figure III-12: Anomalies, decolonizing peace, and peace formation

From this perspective, the involvement of peace agents from outside the local-local realm may be ideally characterized as facilitating local-local expressions, logistically, for instance, or through enabling the invisible to become visible through alternative methodologies. Another form of non-local involvement could be in connecting the different levels of peace to one another by way of networking. Once again, processes are of essence from this perspective.

Complex Adaptive Anomalies

Understanding anomalies as states of complex adaptive systems has allowed us to recognize that sustainability may not be engineered. The connection of the visible and the invisible, through processes that bond different levels of peace may be the set of spheres where decolonizing peace can be expressed as a practice from all stakeholders (Figure III-12). Decolonizing peace emerges in the facilitation of local-local anomalies within their own contexts. Given this perception, a decolonizing answer to Lila Wati's situation, from my perspective, would have been to connect her to Sampat, if she had not already known her. The reality of decolonizing peace is brought to life through interaction and becomes visible through the perceptions of all involved. Should an external agent fail to visualize the invisible, then his or her actions pertain to a dysfunctional category of peace as portrayed in Chapter I. The "when" becomes crucial in trying to highlight pathways toward peace: what is observed becomes part of the observer.

What the Hezbollah and the Gulabi Gang have in common is a structural similarity to the complex adaptive system, which is sustainable by nature. To the questions pertaining to revolution or reform, as panarchy and the al-Qaeda case study have shown, revolt on its own can be as damaging as remembering on its own. Should reform be located within a panarchy, and connected to the resilience of a larger system, then an anomaly may become both a changing and a changed agent for peace. What pathways can be taken to allow for the emergence of complex adaptive systems, not as isolated anomalies but as systemic expressions of decolonizing peace? What pathways can be reached for the linking of different levels of the peace panarchy? The next chapter will look into the relationships that exist between values, ethics, and esthetics in the emergence of decolonizing peace as an inherently sustainable complex adaptive system.

IV. A Journey Through the "Sacred"

As I have been speaking on decolonizing peace across the world, I have been confronted with many types of reactions, ranging from "you are stating the obvious" to "shall we just do nothing and let everyone die?" Often, this pattern of reactions seems to emerge alongside a South/North divide, and the answer to the "shall we let everyone die?" paradigm rarely highlights the systemic Northern involvement in creating, maintaining and nurturing violence, inequalities, and war in what Easterly ironically refers to as the "rest" of the world.[173] In fact, leaving the "rest" alone would be more of a blessing than the liberal peace builders might care to realize.[174] Many authors from subaltern studies or post-colonial studies have deconstructed the neo-colonial approach to North-South intervention in terms of humanitarian assistance, development, or human rights, yet in terms of peace and conflict studies, the post-liberal peace thinkers still escape direct comparisons between their discipline and colonialism in epistemological terms.[175] Wondering how we, the global North, can stand idly by while they, the South, either starve or kill one another not only stems from a *mission civilisatrice* frame, but also from a Northern epistemology that is understood in the global South as abysmal thinking.[176] From this perspective, Northern epistemology is seen as maintaining a system of visible and invisible, making the global South utterly invisible in its agency, and condemning it to recurrently appear as a powerless child, thus shaping the world map into an epistemological cartography of the visible and the invisible. Decolonizing peace seeks to find pathways toward the visible while bridging the North-South Cartesian abyss.

In the summer of 2011, the international community was warned that another famine would affect East Africa, both in Ethiopia and Somalia, as well as other neighboring countries. Visiting Ethiopia at the time, only one hour away from an area undergoing severe food shortage, I was at a loss to understand why food was not an issue in my Addis Ababa capacity-building workshop, given that our neighbors a few kilometers away were facing hunger. Worse, nowhere in the local media was the famine, or food shortages, being referred to, as if Ethiopia had been saved from a systemic regional issue. We decided to limit our food-intake at the coffee breaks, as a sign of respect for the situation that our host country was facing in a parallel reality, since the Ethiopian government always refused to report on it, this while being well aware that our initiative was not going to save anyone. Famine in Ethiopia and Somalia was a chronic issue, popping up on our radar screens once every so often, always within the parameters of the helpless global South needing to be saved from itself. Obscene pictures of naked and starving children reappeared in our media, the visible mantra of a helpless "Other", but nowhere was local agency made visible, much less the systemic, Northern-induced factors that led to the famine.[177] Africans, it seemed, were once again condemned to be pictured as infantile creatures. In fact, the pledge and success of the Somaliland government to organize its own relief effort was almost entirely bypassed by the international as well as local media.[178] It is from this perspective that one can safely ask: when will the global North kick its colonial habit of appearing to benevolently seek to save the "rest" of world? This chapter will seek to develop a decolonizing path of peace and conflict studies from the perspective of cybernetics, looking deeper into the bonding elements of complex adaptive systems. Taking into account the recursive patterns of anomalies as complex adaptive systems, it will seek to provide alternative pathways for decolonizing peace to emerge and nurture itself outside the North-South abysmal thinking and polarization. These bonding pathways will be introduced as a means toward the self-organization of a phase of turbulence.

Addiction

Gregory Bateson has an interesting take on addiction, which was inspired from his observation of the Alcoholics Anonymous' famous "Twelve Steps" program toward recovery. From his perspective, addiction is reinforced by the symmetrical relationship that the addict has with the bottle. The constant state of struggle between the person and his/her addiction, coupled with a sense of Cartesian pride that the self, divorced from the remainder of his/her personality, would be capable of eradicating the bottle alone, would lead the sufferer into a spiral of self-destruction. From Bateson's perspective, the occidental epistemology of dualism lies at the core of the issue. Bateson does not look at addiction as a cause and effect mechanism whose treatment lies in seeking to reinforce the maintenance of sober state, since the tit-for-tat nature of the relationship between the addict and the bottle encourages a lethal escalation. Rather, he looks at the sober life of the alcoholic as driving him/her to drink. From that basis, as a certain "style of sobriety [drove the addict] to drink", he concludes that this "style must contain error or pathology", and therefore, the state of intoxication provides some "correction of this error."[179] As the pathological relationship to addiction, within which a false sense of cause and effect leads the alcoholic into a tit-for-tat escalation with the bottle, the addict "reaches bottom." This is seen as a necessary phase of destruction, within which one realizes that he/she has a serious problem, that he/she is powerless toward alcohol, cannot fix his/her situation alone, and needs professional help. This "reaching bottom" phase is akin to Chapter III's release phase of panarchy. A therapist, combined with a treatment, the help of a community, etc., can initiate this help as part of a reorganization phase, toward an awareness of what Bateson refers to as the complementary relationship that the addict also has with the bottle.

Bateson understands that there can be different patterns of human relationships, which he categorizes as symmetrical or complementary. As explained above, a symmetrical relationship could be defined as an arms race, where both parties react to one another and match each other's behavior in an escalatory fashion. In a complementary relationship, the behaviors of different parties are "*dissimilar* but mutually fit together."[180]

As Bateson explains,

> *"common examples of complementary relationship are dominance-submission, (...) nurturance-dependency, spectator-exhibitionism, and the like."*[181]

For Bateson, pride may lead people to falsely believe that they are the "captain of their soul", and thus identify themselves as alcoholics. Realizing that "the alcoholic" does not begin or end with "the self", however, enables them to realize that the system within which they evolve is part of their addiction. Complementarity enhances the person's scope of being from the incomplete perception of duality to that of a system (see Figure IV-1).

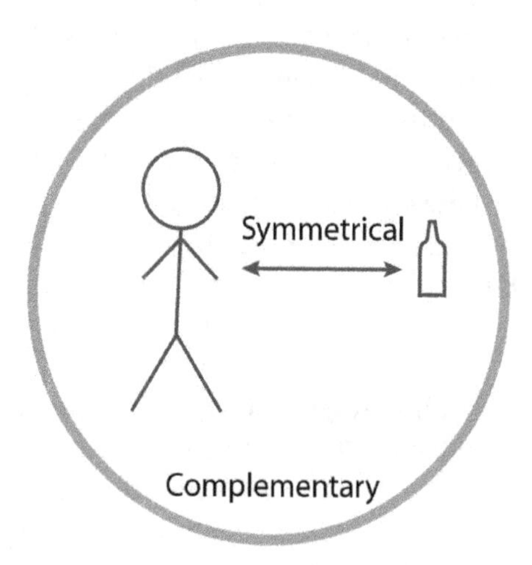

Figure IV-1: Visibility of symmetrical and complementary relationships

As an illustration, Bateson asks what makes a blind person's self in the public sphere: the body alone, or the system that the body-stick-community represents? Where does the characterization of a blind person we see walking in the street lie? With the person? With his/her stick or dog? He concludes that "[t]hese questions are nonsense, because the stick is a pathway along which differences are transmitted under transformation, so that to draw a delimiting line *across* this pathway is to cut off a part of the systemic circuit which determines the blind man's locomotion."[182] What determines the blind person's locomotion is not the conscious act of walking, which does not exist as such, but the "bit" of information that leads the person to walk, i.e. the difference in information that instructs the idle muscles to be in a different state, a walking state. Information from that perspective is not a Cartesian cause and effect mechanism, but a "difference that makes a difference" (see Figure IV-2).[183] In terms of the treatment for addiction, the AA's Twelve Step program and the psychoanalysis related to it seek to enlarge

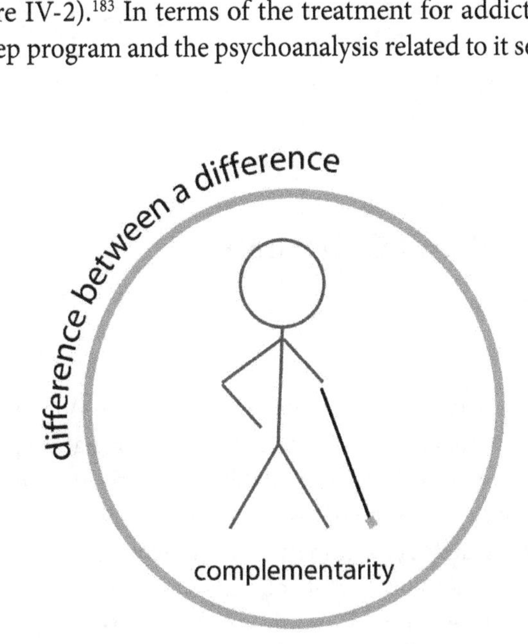

Figure IV-2: Cybernetics of motion

the scope of the addict's symmetrical relationship with the bottle in order to develop an awareness of the complementary, or dialectical, relationship he/she has to it. This awareness represents a difference in the conscientization of one's situation. This difference can be interpreted as a first step toward recovery. In realizing that the addiction will invariably escalate, and that "A's behavior inspires more of B's fitting behavior," the addict will realize that he is powerless within a system that is larger than him/her. Moving from a symmetrical/dualistic perception to a complementary one will be a step forward in the recovery process. In terms of treatment, the therapist, as part of a systemic chain of awareness, will often place the addict in a situation to become aware of this complementarity, such as a heightened state of discomfort that leads to further drinking. Remembering that the action of drinking can be seen as a correction of an error in sobriety, the therapist will enhance this error in order to increase the awareness of the addict. This can take the form of the therapist challenging the alcoholic to partake in "controlled drinking," often resulting in an uncontrollable binge.

Double Bind and Cybernetics

Are all drinkers addicts? Not necessarily of course. What is at issue is not a prohibition but a shift in relationship, or information processing, between all the parts of the system. Bateson remarks that the effects of addiction will be enhanced by the alcoholic pride of the person involved. Additionally, both states of symmetry and complementarity can lead to destruction when they enter a state of schismogenesis, either illustrated by escalation or the emergence of a double bind. A typical complementary relationship can be that of a parent and a child: the child needs the parent's care to develop, but cannot possibly leave this relationship if it is unsatisfactory. Should the parent profess its love to the child, yet be uncaring in its actions, a double bind is formed. An anomaly will emerge within this system, and can take the form of mental illness, dysfunctional behavior, withdrawn attitudes, etc. Bateson argues that "[t]hese potentially pathological developments are due to undamped or uncorrected positive feedback in the system", and that these are "necessarily reduced" in a "mixed system" of both symmetrical and complementary awareness. He concludes: "[t]he armaments race between two nations will be slowed down by acceptance of complementarity themes such as dominance, dependency, admiration, and so forth, between them. It will be speeded up by the repudiation of these themes."[184] The information, or difference within a difference, that stems from this confrontation, will lead down the path to recovery. The awareness of this difference within a difference can be referred to as cybernetics.

Looking at the larger spectrum of addiction in the international sphere, Bateson wrote a seminal article called "From Versailles to Cybernetics," in which he qualifies both the Treaty of Versailles and the discovery of cybernetics as the leading events of the 20th century.[185] In terms of the Treaty of Versailles, Bateson describes how the surrender of the Germans was not obtained as a result of a decisive victory, but through the elaboration of a series of un-kept promises enumerated in President Wilson's Fourteen Points, in which no annexations, no contributions, and no punitive damages were to be asked of the Germans. The drafting of the Treaty of Versailles, renegading on all these promises, would lead to a demoralization of the German people that led to all the major conflicts of the 20th Century.[186] Bateson calls this betrayal an "attitudinal turning point."

What is an attitudinal turning point? Take, once again, the image of a negative feedback loop maintaining a certain set of parameters intact within it. A home thermostat may be understood as a negative feedback loop system. Should the outside temperature be altered, the thermostat within the home will take this difference into account and will regulate the home temperature by either switching the heater or the air-conditioning on or off; this mitigated state in organisms is called the homeostasis of the system. The system works according to a certain set of parameters, which cannot be changed by the system itself: those parameters are called the "bias" of the system. By changing those parameters, i.e. by entering a new target temperature that must be maintained by the system, one will change the "attitude" of the system. From the perspective of cybernetics, the Treaty of Versailles represents an attitudinal turning point in which, according to Bateson, the bias of the international system was altered. This alteration set off a chain reaction of anomalies leading to World War II and the use of the atomic bomb at Hiroshima, itself a recurrent theme in the speeches of Usama bin Laden; the creation of the state of Israel, a recurring theme of the Lebanese Hezbollah, and so on (see Figure IV-3).

Looking at this event at the core of international politics today, leading to conflicts, liberal peace, and economic catastrophes, Bateson states that younger generations, not being aware of this, will find themselves in a state of limbo. He explains: "the people down the line, who were not there at the beginning, find themselves crazy, precisely because they do not know how they got that way."[187] This lack of information, awareness of a difference within a difference, leads to a state of double bind within which younger generations participate in the perpetuation of conflicts that they were not involved in creating. See for instance of how the United states of America legally sanctioned the use of torture after the events of September 11th, 2001. While a signatory of the February 1985 Convention against Torture and Other Cruel, Inhuman, or Degrading Treatment or Punishment, as well as the four Geneva Conventions of 1949, the US, through its Department of Justice, legally restricted the internationally definition of torture to an act specifically intended to cause "organ failure or death."[188]

The impossibility for current generations to reflect on an earlier attitudinal change has created a situation in which torture is legally condoned by a Western "democracy", and its subsequent use in secret prisons, the infamous rendition programs and Guantanamo Bay has been used by many insurgencies across the world as an excuse for politically violent dissent and the perpetration of acts of terrorism. In this context, Bateson introduces the concept of cybernetics as the second major occurrence of the 20th century. Cybernetics, the understanding of information as a difference within a difference within a larger system, enables us to enter into a complementary relationship with the rest of the world. By becoming aware of the double bind within which we find ourselves, the potential for a paradigm shift of our international system will undoubtedly be heightened.

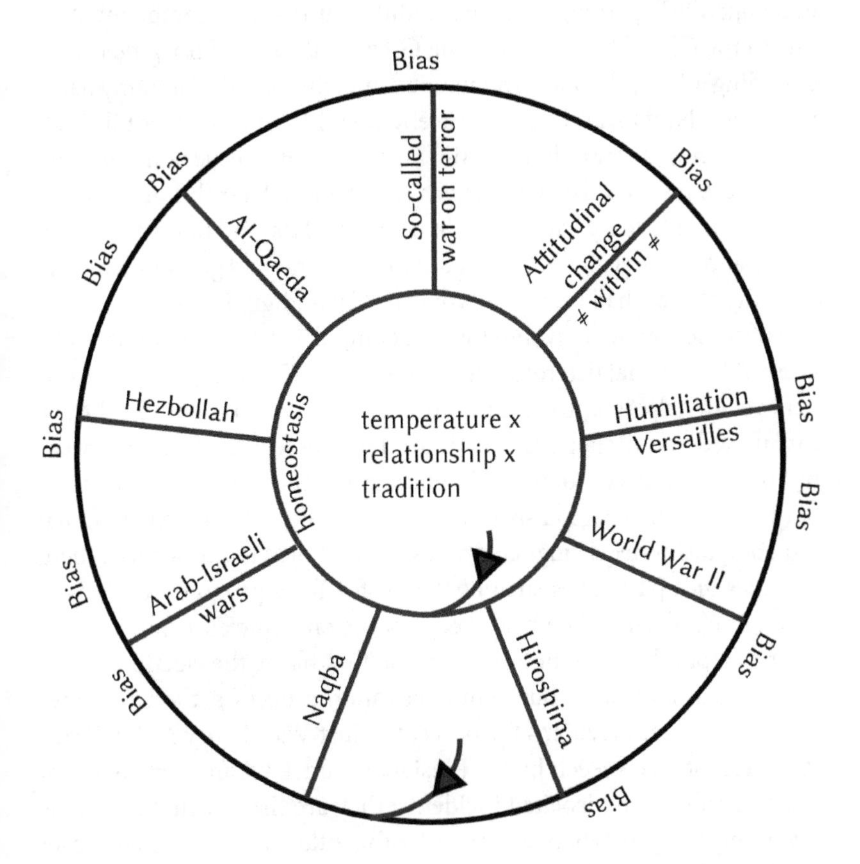

Figure IV-3: One attitudinal change invisible to future generations with subsequent conflict

Cybernetics of Peace

From the perspective of addiction and cybernetics, the "shall we let everyone die?" paradigm reaches a different realm, a complementary one to the Global South. Honoring Gregory Bateson, I imagine myself attending a "Peacebuilder Anonymous" meeting and declare: "my name is Victoria Fontan, and I am a peacebuilder. I tried to shower Lila Wati with money as a way to both gratify my benevolent image of my Self and to appease my troubled Westerner's conscience. After this act of kindness, I was able to fly back home in business class, and never to have to worry about her again, because I did my best and had good intentions." Looking at the "shall we let everyone die?" paradigm from the perspective of addiction, we may find that the compulsion to "help" unilaterally is an addiction that destroys our surroundings. Our drunkenness is our benevolent help: it represents the correction of the error of our sobriety, manifested in our realpolitik. Of importance here is the fact that we are not necessarily aware of the difference within the difference that led us to this situation of compulsive and destructive help. The destruction that our compulsive help brings around us is manifested in tip-of-the-iceberg scandals and paradoxes such as the oil-for-food program, the sexual abuse of children by UN peacekeepers, the endemic corruption of the interlocutors that we empowered in the first place, the election of war criminals as a result of our democratization processes, the US drones that kill innocent civilians for peace, the Just War theory Nobel Peace Prize acceptance speech by US President Barack Obama, etc. None of these paradoxes are isolated incidents; they are the destruction that is invariably brought when we are under the influence of the "shall we let everyone die?" paradigm.

A cybernetics of peace brings to our visibility the differences within differences that have caused the catastrophes of our well intentioned peace throughout the "rest" of the world. While many publications of the peace and conflict studies field focus on how we should help, train, or empower the "other" into being more peaceful, always choosing which paradigm to develop for them, which theory to train them on, etc., I argue that the first step toward decolonizing peace and conflict studies must be our realization that we are powerless in facing the destruction and injustices that we invariably bring with our peace.

While not every person that drinks will become an alcoholic, the contention that "once an alcoholic, always an alcoholic" ought to prevail in the sense that no matter how many "best practices" guides are going to be written, no matter how much we try to "do no harm", the parameters of our dysfunctional symmetrical relationship to peace and our "subjects" will never change, because when looking at this relationship from a complementary perspective, we are always in power and always at the higher level of the panarchy, not morally, ethnically, or racially, but politically and economically (for now). We call the shots, rhetorically and in practice at the UN Security Council, for instance, and therefore ought to realize that we will never be in an equal relationship with the "rest" of the world, unless they find a way to turn the tables from a power perspective. A cybernetics of peace is therefore a pathway toward the visibility of our own addiction, the differences within the differences that have led to today's conflict scenarios across the world, and the realization that we cannot trust ourselves with our "help", in the same way that the alcoholic will never be able to engage in controlled drinking. Decolonizing peace and conflict studies activates a corrected positive feedback into the system in terms of raising our awareness of the existence of a mixed symmetrical and complementary relationship to "help." The question is therefore not "shall we let everyone die?" but "what can we do to facilitate alternatives, since we cannot trust ourselves with the help/peace we bring?" This represents a paradigm shift that bonds us to the "rest."

Awareness of our position in a higher level of the panarchy of peace allows us to understand where we can best be positioned in our possible remember connection to the lower levels, as well as their potential for dampening ours with necessary positive feedback loops, thus reducing our risk of developing into the rigidity trap. Should this not be facilitated, a potential for schismogenesis materializes with the development of anomalies, which emerge and are treated as an obstruction between us and the lower levels. This was the case of the Hezbollah in its early years, which led to a sudden surge in violence. Anomalies cannot be avoided, yet their effects depend on our awareness of ourselves, within a mixed system of symmetry and complementarity. A necessity for another attitudinal change or paradigm shift is therefore crucial if we, in peace and conflict studies, choose to remain a trusted interlocutor with the "rest" of the world, excluding of course our trusted interlocutors' elite, who serve us well yet maintain structural violence for the "rest". Of importance here is that a parallel dimension is present, working, and successful, as illustrated with Sampat's experience. It does not need us to thrive; it is powerful in its own right and expression of agency. Can there be a mutually beneficial link between all levels of the panarchy of peace (outlined in Chapter III), provided that the dysfunction of the "help" addiction is kept under control? Initiatives such as that of Sampat are legion, as the use of an alternative set of epistemological lenses allows us to realize. As they make a difference, they are also, of course, able to make a difference to us in our own environments, through a revolt connection. The Arab Spring illustrates this in relation to the inspiration it gave in the formation of the Occupy movement as a turbulence.

From Tahrir to Wall Street

Many comments have been made on the links that exist between the events of the Arab Spring and other mass protest movements across the world, culminating in visibility in the Occupy Wall Street movement of 2011-2012.[189] While it was recognized in Chapter II that the Arab Spring did not bring lasting change to the power dynamics within the region, and that the revolution that was hoped for resulted in mere a re-shuffling of the cards, has the spread of discontent across the world not set a difference in relation to the expression of public dissent? Has the success of the occupation of Tahrir Square, and the illusion of the removal of dictatorship not encouraged others to occupy their own centers of power and inequalities? It is undeniable, in that sense, that the Arab Spring has provided a revolt connection toward the higher level of the peace panarchy; provoking us, the greater North, to question the foundations of our own socio-economic system. Of importance to this revolt connection is that it has provided a difference within a difference that has pitted us against our own injustices both in content and form. In terms of form, the organic nature of the complex adaptive system can actually be seen. Michael Marder, for example, understands the occupy movement as a vegetal occupation that no longer seeks to exist within time or rational-thinking, crossing the boundary between the immanent and the transcendent image of democracy, but rather, to mimic nature in its non-violent yet visible and unavoidable occupation of a physical space that lies at the center of the public sphere. Visibility through plant-like occupation represents the most encouraging anomaly toward a paradigm shift in social change today. Marder explains:

> *"[p]urely vegetal beings do not protest, do not set themselves against anything, do not negate—symbolically or otherwise—what is. But if we act as though we were them, (...) we would, consequently, repudiate the ideal of sovereign and decisive action, directed by a rational, conscious or self-conscious, individual or collective subject and, instead, nurture the horizontally and anarchically growing grassroots that crop up wherever protest tents are pitched in the shadow of skyscrapers."*[190]

While the Occupy movement has been deemed by its critics to be aimless, scattered, and without a clear set of demands, the common vision that it presents is one of visibility, that of the 99%. It does not exist in the form of a traditional political manifesto, yet the processes that it uses are clearly akin to the elements that nurture strange attractors in terms of shared vision, team processes, and information flow. Those elements, integral to complex adaptive systems, are by essence organic.

Within the Occupy movement, the decision-making process in itself challenges traditional methods, since all decisions are taken on a consensual basis. Meetings, conducted in a circle, allow everyone present to speak, and are collectively steered by a set of three different gestures that signify agreement, disagreement, or veto. When someone places a veto, this person is then allowed to explain the extent of his or her disapproval, and the meeting continues until a consensus has been established.[191] This is a rather different process, which can be as benign as it can be damaging outside an actual paradigm shift. A consensus by name only, the map to what the territory is not, can indeed lead to frustrating experiences, the most frustrating of all, according to Paulina Gonzalez, being the appropriation of the 99% movement in the case of Occupy Los Angeles by a crowd of "help" addicts who ended up marginalizing the lower social segments of the movement.[192] In this particular instance, hotel workers were not taken into account and respected by the *bourgeois-bohemian* lot of Occupy LA, who had decided that since their union was not showing up at Occupy meetings, the migrant workers who nonetheless were present were not going to be given any formal space in the movement. From this perspective, one can see that an organic process can also be highjacked by the greatest threat to peace, the benevolent peacebuilder that decides what is best for the "rest". Still, the visibility of the issues at stake and the entire Occupy process have been unaltered, as the resilience of the system has allowed other Occupy movements across the USA to thrive. While the greatest critique of the Occupy movement from a Southern perspective has been, "where were you all along when we were being dispossessed by you for your own gluttonous economic growth," of interest here are the processes that such an alternative, initiated by a revolt connection, may represent in relation to decolonizing peace.[193] Undeniably, for the revolt connection to be successful as a positive feedback loop, an attitudinal change remains crucial. It is my view that the Occupy movement will only be successful in the North if its "help" addiction is faced in a mixed symmetrical and complementary manner. The attitudinal change remains in this awareness.

Dr. Jekyll and Mr. Hyde

Gregory Bateson offers a suggestion for bridging the North and the South, for mixing symmetry and complementarity within the establishment of revolt and remember connections, in what he refers to as the sacred. Not to worry, there is no subliminal religious message to this book, nor is there any call for peace spirituality; I leave this to benevolent "helpers."[194] Bateson's notion of the sacred is the connection between the left and right sides of the human brain, which respectively account for rational/Cartesian prose-type of thinking, and the dream/poetry of the now moment (Fig. IV-4). Brain scientist Jill Bolte Taylor explains it as: "[w]hen normally connected, the two hemispheres complement and enhance one another's abilities. When surgically separated, the two hemispheres function as two independent brains with unique personalities, often described as the Dr. Jekyll and Mr. Hyde."[195] The left side of the brain is responsible for our sequential understanding of where we are, who we are, and what we are thinking. It is connected to the endless chatterbox that speaks in our heads, and it is also the host hemisphere of the ego. The left hemisphere relates us in time, according to a separation of the past, present, and future. It relates to our environment in a rational and analytical manner, dissects all parts of what we see and experience. As Bolte Taylor explains:

> *"[o]ur left hemisphere looks at a flower and names the different parts making up the whole –the petal, stem, stamen, and pollen.(...) It thrives on weaving facts and details into a story."*[196]

From this perspective, it is not difficult to understand how the liberal peace of the "helper" comes about as a strict left hemisphere expression of peace. Of importance here is to realize, even though it may seem cliché, that peace is more than the sum of its parts, in the same way that we as organic human beings are more than our left hemisphere.

The different expressions and definitions of liberal peace exposed by Richmond– internal/external binary, hegemonic act, bottom-up or top-down construction, temporal, geographical, leveled, etc. – all pertain to our left hemispheric understanding of reality.[197] It does not mean that these are wrong, it means that they are incomplete, and that within this limitation, they can become lethal to both the individuals and the communities that they find themselves entitled to be "helping."

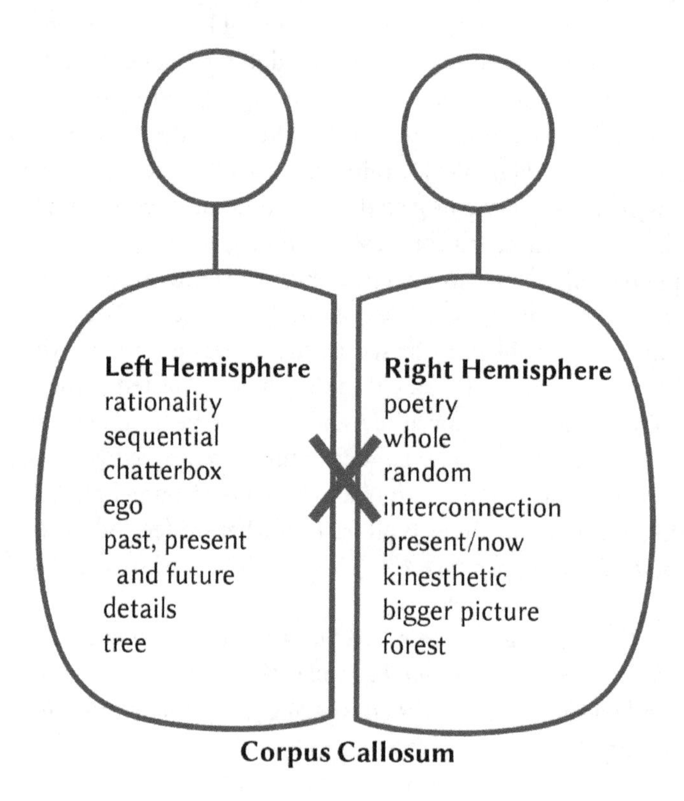

Figure IV-4: Dr. Jekyll and Mr. Hyde

The right hemisphere of the brain, taken alone as a separate entity or personality, is just the complementary opposite of its left counterpart. As Bolte Taylor explains: [o]ur right hemisphere is designed to remember things as they relate to one another. Borders between specific entities are softened, and complex mental collages can be recalled in their entirety as combinations of images, kinesthetic and physiology."[198]

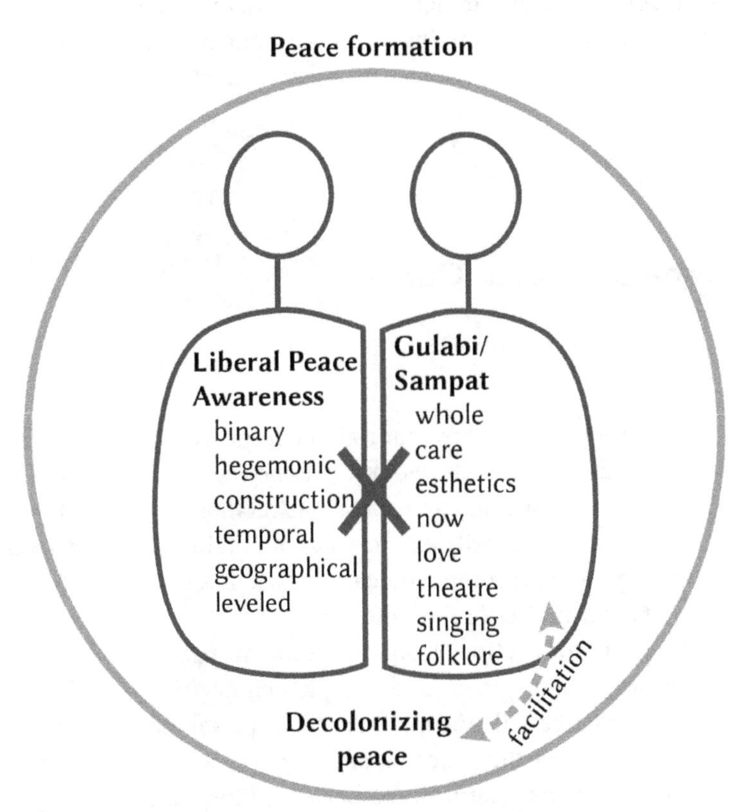

Figure IV-5: Peace seen through the hemispheric lens

The right hemisphere is connected to the present moment; "it perceives the big picture, how everything is related, and how we all join together to make up the whole." More importantly, "[t]he present moment is a time when everything and everyone are connected together as *one*."[199] The instinctual practice of peace of Sampat lies in a right hemispheric realm. Her appeal to songs, theatre, and local folklore reflects a kinesthetic understanding of peace. Sampat does not worry about who did what before or after her in relation to her given situation and that of her peers, which is why in spite of much courtship on behalf of political parties, she was never recuperated politically. The future and what it might be does not matter to her, the present moment and her connection to all around her prevails in her thinking and actions. She acts as a bonding connector of the whole, within her own environment, period (see Figure IV-5).

Bateson's "Sacred" Curse?

Since "[j]ust opposite to how our right hemisphere thinks in pictures and perceives the bigger picture of the present moment, our left mind thrives on details, details, and more details about those details," a decolonizing approach to peace will seek to act as our brain's corpus callosum in connecting the "two complementary halves of a whole rather than as two individual entities or identities."[200]

This "sacred" connection, from a Batesonian perspective, mixing symmetrical/left hemisphere and complementary/right hemisphere relationships to peace, will not only balance the power relations between the different levels of the peace panarchy, but it will also provide an understanding of the relationships between kinesthetic and logical understandings of peace (Figure IV-5). Connection comes as the facilitation that was introduced in previous chapters, through the process of bonding. Does this mean "bringing" more of the right hemisphere into a left hemisphere understanding of peace? Absolutely not; as previously established, Sampat-like initiatives cannot be engineered. Yes, you might have already guessed, my first aim at meeting Sampat was to see how her model could be replicated, let's say, in Afghanistan, where I as a benevolent feminist had decided that Sampat was just what Afghanistan needed. After I disclosed my motives to her, Sampat laughed it off, and said that she did not want to be responsible for anyone's death… only her own. No, we definitely cannot be trusted with our "help."

What remains of essence is connection and facilitation as valuable steps toward decolonizing peace, not engineering it as a decolonized end. While Bolte Taylor advocates for a "stepping to the right" side of our left brains to experience greater inner peace, Bateson warns against the recuperation of the sacred for commercial, and dare I add, political gains. He explains: "you've got something nice, central to our civilization, which bonds together all sorts of values connected with love, hate, pain, joy, and the rest, a fantastic bridging synthesis, a way to make life make a certain sort of sense. And the next thing is that people use that sacred bridge in order to sell things."[201] The artistic exposure of starving children calls on our sacred sense of connection to the whole, yet it serves to stuff the economic resources of the peace industry, with only a ridiculously small percentage of our donations reaching the "field."[202] The exposure of massacres in Libya compelled us to give our blessing for the military removal of a dictator, yet it served the interests of Nicolas Sarkozy in his desperate attempts to flog his *Rafale* airplane to foreign military forces.[203] The use of the sacred might sell Benetton clothes, but it also sells war, destruction, and rape, all under the guise of help.[204] The artificial practice of the sacred is what makes the "shall we let everyone die?" argument so compelling, irrationally appealing to our right hemispheric sense of the whole, yet invariably precipitating in the dangerous consequences exposed in Chapter I. As Bateson once wrote, the "sacred" is misused.

If so, why suggest the development of a sense of the sacred in this book? Will the discussion of decolonizing peace, complex adaptive systems, the "help" addiction, or the peace panarchy, add in some way to the vulnerability of a community facing the Peace Empire? I have been unable to answer this question for some time, until I came to the realization that a mere critique and deconstruction of the liberal peace paradigm would never been given any importance if alternatives were not portrayed, analyzed, and presented as a viable parallel reality to the false choice imposed by the "shall we let everyone die?" paradigm. The "sacred" as the connection between an addiction-aware left hemispheric peace, and a complex adaptive system/Sampat-like right hemisphere, remains one of the most viable paths toward decolonizing peace. The visibility of the invisible is a crucial part of this equation, as the visibilised invisible provides the bonding difference within the difference that both informs us of viable alternatives as well as places liberal peace within an awareness of its "help" addiction. Visibility in that sense is unavoidable, since it connects all within the realm of Richmond's peace formation (Figure IV5). True, skeptics will always refer to the visibilised invisible as an exception, an anomaly, yet from a decolonizing epistemological standpoint, it is undeniable that the visibilised invisible is more than an exception, it is a world that already exists. Can there be a bridge between both the Cartesian visible left hemisphere and the visibilised invisible right hemisphere?

Kinesthetics of Peace

Can the existence of common values act as the "sacred" bridge between the left and right hemispheres of peace? Those values may share the same names, but their interpretations might be radically different. Since many crimes against humanity have been committed in the name of peace, its invocation may not mean anything to many. Should certain values of the right hemisphere be privileged, since the left cannot be trusted with them? Critical pedagogy and peace education assert that a culture of peace is to be established through the nurturing of all the different learning capacities of human beings.[205] From this perspective, the kinesthetic mind uses the body to create, or do. We also understand from Chapter I that since the map is not the territory, what happens in the territory is the reality to be reckoned with. If actions speak louder than words, then the kinesthetics of peace are found in the right hemisphere expressions of peace, in Sampat's living of peace, in the bonding elements of the complex adaptive system that the Gulabi Gang represents. It has already been established that she relies on esthetics in her symbolism and recruitment mechanisms. Ethics can also be seen at the very core of her action and that of her group, as well as their claims toward the government and social systems within which they evolve. One key element of the kinesthetic of peace that remains to be deepened is the practice of care. What if care as a shared value, yet practiced in the complex adaptive system, could act as a bridge? Leonardo Boff writes that *"[t]o care is more than a mere act; it is rather an attitude."*[206] I contend that care, alongside love, which is recurrent in Sampat's discourse, is part of the attitudinal change that will alter the bias of our discipline and its practice.

Care encompasses many aspects of both human life and its relationship to nature, culminating in what Bateson refers to as the ecology of the mind. Care, as part of the Gulabi gang, is manifested in the open circle of help: I care for you if/and you will care for someone else. One other group in particular has caught my attention in terms of care, and will come as a direct answer to Richmond's set of questions in the concluding chapter of *The Transformation of Peace*: "what can the (...) peacebuilding official (...) do when, on a UN helicopter taking off after a meeting in a rebel held village in Eastern Congo, a woman tries to put a sick child on board so it can receive desperately needed medical care in a faraway town?"[207]

In 1991, as Somalia was being destroyed by a civil war, Hiro Adam Diriye Samawada was fleeing a bombardment in the country's capital, Mogadishu.[208] She happened to hear a baby crying in a nearby house, and as she approached, she realized that the baby's family was dead, and that it would probably be left to die alone if she did not intervene. As she recalls this event, she tells me that she did not think too much – right hemispheric thinking and bond to humanity – her life was in danger too, so she just took the child in her arms and ran away. It was a boy. Since the Somali clan system is tightly knit, she thought that she would be able to find the boy's family and drop him off with an uncle or a relative, but as the days went by, she could not find anyone to entrust him to. Everyone was either dead or had fled that part of town. After a few weeks, she resolved to care for him herself. As the times were harsh, and the war presented Somalia with an unprecedented situation both politically and socially, it was not unusual for anyone, even a widow, to care for other people's children. As Somalia was in a release phase of its panarchy complex adaptive cycle, desperate times called for desperate measures. Born in a wealthy family in nearby Djibouti, and separated from a husband that lived abroad, Samawada needed not worry about the financial or familial aspects of her decision. A few months later, a similar situation occurred; this time it was a little girl that she found. She had the same reaction as before, and over the years, her family has grown exponentially. On the day I met her in Hargeisha, Somaliland, in November 2009, she had more than 100 children under her care, and could not exactly remember how many children she had been caring for until then. So where exactly had this urge to care come from?

People Are the Best Resources We Have

She says that the turning point of her life occurred when she visited a jail in Ethiopia in 1978. As a lawyer, she had come to follow-up on the case of a fellow countryman when she heard a scream. As she discreetly moved to the scene of the shouting, she saw a man being tortured. She stood there, paralyzed by the horror that was taking place before her eyes: a piece wood was being nailed with a hammer alongside the man's leg. The pain it caused was unbearable, and even after the man passed out, she stood there, aghast, terrified. She recalls that she knew about torture, but that she had never witnessed it first hand. The man in question was accused of being a smuggler, trying to feed his family, so from that day on, she decided to look after the legal rights of the poor. All the money she had was to be spent helping detainees awaiting trial to avoid the torture that routinely pertained to the poorest of prisoners. As a civil war broke off in Somalia and its legal system collapsed, her activities came to a near halt, until she found her first child.

Now based in Somaliland, she organizes a network of care for her 100 children. While the war is over, and Somaliland has brokered its own peace in the 1990s, social traditions still reject the conception of children outside marriage, and as a result, all babies that are conceived that way are abandoned in the outskirts of the cities, left under trees until hyenas come to eat them alive. Samawada's job is to pick them up before it is too late, and to care for them until they are able to look after themselves. Her initiative is now well known, with paintings symbolizing her actions scattered around her district of Hargeisa, the capital of Somaliland. Those paintings represent a baby laid under a thorn tree in the desert, with Samawada's contact details underneath. When someone knows of a case, he or she can just alert her, and she or a relative will come and pick the baby up. The children are now spread over different houses across her neighborhood, and she runs many different schools, some under trees, and some in actual buildings. She applies her legal skills to adopting all her children and, as she explains, redistributing them in various clans. This does not mean that these clans physically look after the children, but it does provide them with a future, since in Somalia and Somaliland, it is not possible to exist outside of a tightly-knit clan structure. In order to feed her family, she relies on the help of her extended network of friends and the children she has looked after. Some families donate property to her, others spend time with the youngest, cook food for the children housed nearby, etc. Needless to say, she relies on no donations from international organizations or international NGOs. Once, World Vision visited her. They resolved to build a concrete floor for her school, took a picture of happy children in the classroom, and never returned. Samawada was told that it was for their annual calendar. Indeed, there is no such thing as a free lunch when it comes to benevolent "help."

Understanding that if she had waited for anyone's "help", she would never have picked up that baby so many years ago, Samawada created her own clan that is now self-sufficient. Care, in her eyes, is the most important aspect of her work. As we spoke, she repeatedly stated her conviction that people are the "best resource that we have." When I asked her if she would care for funding, she said yes, but then retracted and told me that she needs to be very cautious, that too much money could come to harm her carefully weaved network. A very close acquaintance of hers, Edna Adam Ismael, made an issue of money and somehow, according to Samawada, this broke the close ties that they once had.

Edna Adam Ismael is a remarkable woman.[209] The daughter of a prominent medical Doctor, Adan Ismail, she grew up in an affluent environment. Being a woman, she soon realized that her place within her traditional society would not be where she might want it to be, but she was resolved to carve a space for herself with her country's public sphere.[210] At the end of the 1950s, she undertook nursing studies in England, and in order to be able to commute to and from her university, she also learned how to drive. Occasionally, as her dad visited London, she would offer him rides across town. Upon finishing her studies, in 1961, she returned home to Somalia and was resolved to keep driving. Back then, there was no law in Somaliland that prohibited women from driving, yet from a traditional perspective, it was unthinkable. She decided nonetheless to keep driving and offered a ride to her dad. They were immediately stopped by the police, who took the car keys away from Edna, and argued that she should not be driving. Thankfully, they had kept a spare key, and repeated their operation the next day. This time, the police offered to have one of them drive Edna around whenever she wanted to go out. Dr. Ismail responded that he would never trust his daughter, nor his car, to any of their drivers, and that she would continue to drive. After they realized that Edna did not have a Somali driver's license, they requested that she take a driving test. Six months later, after taking six driving tests, since no one could believe that she could actually do it correctly and kept asking for additional aptitude proofs, she was reluctantly given a Somali Drivers License, that was in February 1962. She effectively became the first woman driver in the history of her country. That was the first of many bifurcation points in Edna's life. In relation to chaos theory, Edna Adan is a trickster, an agent of radical transformation within her society.[211]

Her elite status married her off well, and she later became Somalia's first lady, as well as Somaliland's Foreign Minister between 2003 and 2006. As a trained nurse, she opened her maternity hospital in 2002, where she is training midwives to help alleviate one of the highest maternal mortality rates in the world. The international community has widely praised her courageous actions; she has been featured in Nicolas Kristoff and Sheryl WuDunn's *Half the Sky* New York Times best-seller, and she has been able to rely on a steady stream of Northern midwives to come and assist in her hospital.[212] Her journey to help transform her society has been flawless, irreproachable, and courageous. She used all her savings to build this hospital, and has never ceased to care for and inspire all the women that she has trained. She receives and hosts all the internationals that come to visit her in Hargeisa and seek to understand the situation on the "ground." She spends half her life in between airports, living out of her suitcase, and is now in partnership with important donors such as USAID, which help her keep the hospital open.

Connecting Two Worlds In the Same Town

There is no "but"; Edna is helping her community according to her status, vision, and networking abilities. Yet when one of Samawada's children needs medical attention, the fact that she cannot afford the fees means that the child will not receive treatment from Edna's hospital. This has caused Samawada great anxiety, and has led to the death of some of her children. She does not understand why, after putting her wealth to the service of her children, comparatively "rich" hospitals such as Edna's could not give back to the community by treating orphans for free. Samawada could go to another hospital, but Edna's is the best in the area, and she asserts that all the "good" doctors work there, at the expense of other medical facilities. Moreover, should a child see a doctor, he or she would not be able to give adequate medication free of charge. From Edna's perspective, she has to run a hospital that is always on the brink of bankruptcy, and she cannot be seen to offer services or medicine for free. Can there be any connection between the left and the right hemispheres of peace in this particular case, a connection that would be meaningful to all, without perverting either? What form would this connection take, since the "help" of NGOs, foreign governments, and International Organizations cannot be trusted? It is clear that a free caring for the medical needs of Samawada's children would extend her own network of care, facilitate her daily actions, and keep the children alive and well. Could creative local-local solutions be found to link both initiatives? It is not for me or anyone outside this system to say; yet one can imagine how win-win solutions might be found to resolve this particular issue, to help enhancing the network of care that is already part of this vibrant community of practice.

Money is not the solution. The local-local elaboration of partnerships may create the necessary link when a remember connection from the above level of the peace panarchy interacts with the lower level in a relevant way. The liberal peace model fully utilizes Edna; she is part of the higher level of the peace panarchy, yet embodies hybridity. She represents the trusted local interlocutor who speaks our language, uses our peacebuilding keywords, indirectly appeals to our flawed sense of the sacred – it is not her, it is us – and makes us comfortable as donors that our money will be well spent by one of us, the local, trusted elite, educated by us. What she represents to us allows us to drop our golden coin with great pomp so that our conscience may be appeased, so that we may be assured that we did not "let them die" in the end. Yet Edna is also a sincere and unbroken free spirit who utilizes the liberal peace industry to make a difference at home. She does not live in Geneva, does not fly business class to deliver graduation addresses that infantilize the great suffering of her African sisters. She lives in her hospital, with her people in Hargeisa, and when at home, is on call at all hours of the day.

She has continued delivering babies herself since her hospital opened. Even though some of her detractors at home deplore the fact that she only employs staff from her clan, who in the Somaliland elite does any different? Which Member of Parliament can claim to employ close bodyguards from other clans?

The remember connection between Edna's and Samawada's levels of the peace panarchy may be fostered by the creative, local brokering of a connection between both. From that perspective, Samawada's right hemispheric care can be nurtured by Edna's hybrid left hemispheric utilization of the liberal peace paradigm. Samawada could be the revolt connection that enhances the scope of the care-based network, and Edna could be the recipient of the remember connection from a higher level that is aware of its "help" addiction (Figure IV-6). This awareness, again, is crucial, as some liberal wolves can wear hybrid clothes. The realm of the Samawada revolt connection to Edna as the hybrid, and the remember connection coming to Edna from above represents the scope of Bateson's "sacred." For it to be successful, it needs to happen on their own terms, and on their own account. Hybridity is the missing link between the aware post-liberal peacebuilder and the Sampat-type local-local, while the "sacred" lies in the connection of the hybrid with the local-local. As I mention this possibility to Samawada, she tells me that she will be thinking about it, that she is not sure that it can work, but that she will try to see. Our conversation is then cut short by the "Mission Impossible" ring tone of her mobile phone… another baby is waiting for her under a thorn tree.

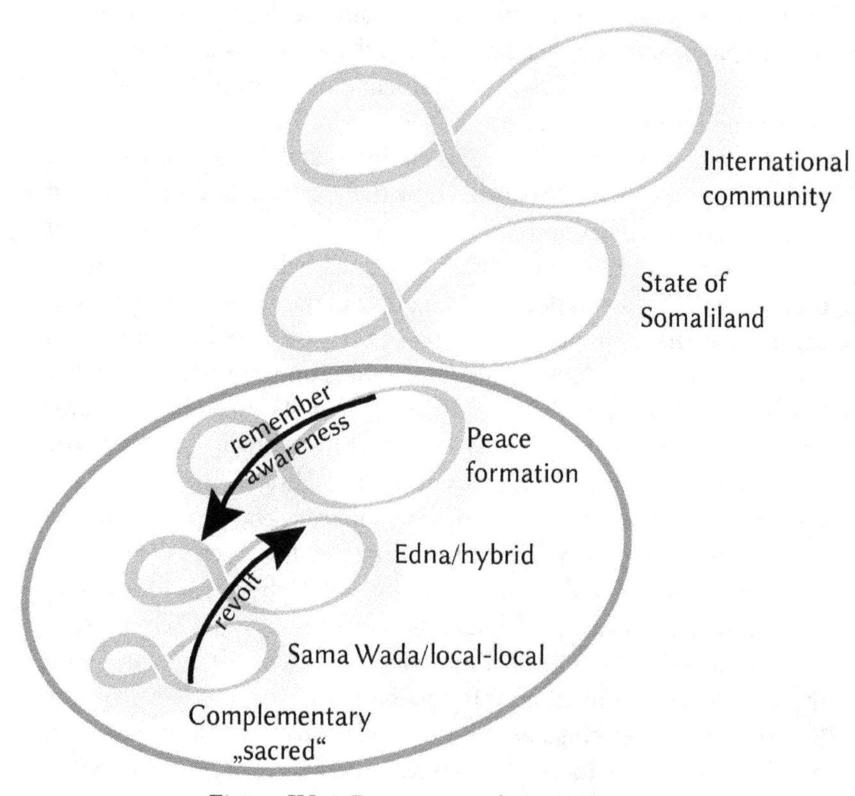

Figure IV-6: Peace panarchy in Somaliland

Conclusion

In answer to the question posed earlier, pertaining to the baby child being taken or not into the UN helicopter, the visibility of the invisible may provide a viable alternative to the removal of the child from his/her environment. Visibility and awareness within a greater peace formation context is the key to approaching the complexity of the situations that are encountered both in the "field" and at home, as illustrated by the lessons learned of Occupy LA. From a Northern liberal perspective, this chapter has sought to bring to our understanding the necessity of treating our "help" syndrome within a mix of complementary and symmetrical awareness. While we cannot be trusted with our "help", the facilitation and reliance on hybrid actors such as Edna Adam may be the key to a "sacred" facilitation of the values and bonding elements that are found in complex adaptive systems such as that of her friend Samawada. Part of this sacred is the practice of bonding values such as love and care, as well as a kinesthetic understanding of peace, relying on ethics, esthetics, self-organization, consensus, etc.

Decolonizing peace does not depend on tools, but on processes and actions. Can there be some elements that facilitate the emergence and amplification of self-organized complex adaptive systems, which bring a common vision to the realm of the visible, etc? Can the close study of several amplifying aspects of the expression of decolonizing peace bring about an understanding of its processes? The last chapter of this book will look more closely at some of the elements that facilitate the emergence and amplification of turbulences and complex adaptive systems through networking, leadership, and dialogue.

V. Emergence

In March 2011, an interesting difference occurred in my academic life. I carried out "field" research from my home base, comfortably sitting behind my desk at work, saving myself from the despairing boredom of faculty meetings, or waiting for my daughter to fall asleep at night. I spent weeks observing Internet Relay Chats (IRCs) of Anonymous (Anon). It all started fairly simply, a friend of a friend of a friend knew someone who was remotely connected to the IRC channel of Operation Iran (#OpIran), and he connected us. Anonymous had made itself known to the rest of the world in its connection to the Egyptian uprising, and I was curious to understand their functioning, organization, political remits, etc.[213] I received a message in my mailbox asking me to submit some of my previous publications, so that some of the people who I would be interviewing could judge my credentials before they would decide whether or not to relate to me. Three academic papers of mine were uploaded by my contact onto Scribner, including an early draft of this book's chapter 1. Over the following days, they received significant attention. After a deliberation among Anon members of Operation Iran (#OpIran) and beyond, Operation Teach (#OpTeach) was established, followed a few days later by Operation Learn (#OpLearn). To get onto #OpTeach was no easy task. I had to follow many technical instructions, but more importantly, surrender to my contact, someone I did not know personally at all. Through the links that I clicked on and the programs I downloaded, any hacker could have taken control of my life; they knew I was working on a Mac, for instance, what software I was using, where I was located, etc.[214]

Trust, from this perspective, was the most important part of this research. As a bonding element of a complex adaptive system, it connects to the right hemispheric understanding of being part of a whole. A bonding element between my contact and me was necessary for me to look deeper into the amplifying elements of Anonymous, one of the complex adaptive systems getting much media attention.

Decolonizing peace has so far been understood in this book to be bringing the invisible into the realm of the visible. Whether it be in relation to presenting an alternative epistemology of the discipline of peace and conflict studies, to facilitating the connection of the left and right hemispheric expressions of peace, to highlighting the period in the panarchy cycle of peace when outside involvement/facilitation might become valid, to understanding the emergence of sustainability in complex adaptive systems, to grasping the double bind of liberal peace, or to highlighting the importance of hybridity for revolt and remember connections in relation to the peace panarchy, decolonizing peace has sought to add a territorial complexity onto the mapping of post-liberal peacebuilding[215]. While decolonizing peace emerges as a complex expression within the greater realm of peace formation, both complement one another in the peace panarchy.

Now that the connections have been established within the understanding that decolonizing peace can be facilitated along a kinesthetic path, neither built nor engineered, it remains necessary to look deeper into the factors that enable the negative feedback dampening of a system, as well as the amplification of a dampened closed system into both a turbulence and a vortex, as well as beyond. What factors will enable the formation of positive feedback loops? This last chapter will look closely at the emergence of networks, the alternative expression of leadership within them, their self-organized elements, and the dialogue processes that connect all of their parts and strengthen their resilience. The importance of networks, communities of practice andsystems of influence will be analyzed as a decolonizing expression of both the self and the collective.

From Networks to Communities of Practice

We have already seen from Chapter III how the aggregation of negative and positive feedback loops comes to a phase of bifurcation and amplification, the turbulence moving toward the shaping of a vortex whose parameters are symbolized by the emergence and nurturing of strange attractors, which provide boundaries as well as a space for unpredictable movements. From that perspective, the pattern of a vortex remains the same, while its details change; it renews itself constantly. Think about the human body and the fact that our organs remain present, but the cells within them are constantly renewing. Our liver remains a liver, in its pattern, yet its composition is undergoing constant change. In nature, and especially in my garden, this is also illustrated in colonies of *zompopas*, Leafcutter ants, who maintain an ability to eat a small tree in one night, over and again, all year long, while the life expectancy of each individual ants is between one and two months.[216] Fritjof Capra argues that the same happens for live networks and organizations whose boundaries are clearly established, yet whose membership can be renewed constantly. For the renewal of the cells to be optimized, Capra emphasizes the importance of the constant creation of knowledge within the network, as well as a constant flow of information between different members.

From a Quantum Communications concept, relationships between network members make matter, not the members themselves. Networks need membership, but their resilience is not due to numbers or a critical mass, rather, it relies on the relationships between the members. This should not be confused with the experience of Samawada in Chapter IV, who clearly stated that people were the best resources that her network had. What I believe she meant in her particular context is that people, as opposed to funding, and the relationships that they have established between one another through the practice of care, and love of course, provide the drive behind her network. From this perspective, the "cells" of the network are the different nodes that compose it, either individuals or groups of individuals, and the energy of the network comes from the relationships between the nodes, illustrated by the creation of knowledge and the flows of information. Energy is equally present in all these nodes, and does not belong to any central "regulating body." From this perspective of decentralization, the constant flow of energy through information and knowledge creates the self-organized aspect of the live network. There are no hierarchies, only nodes.

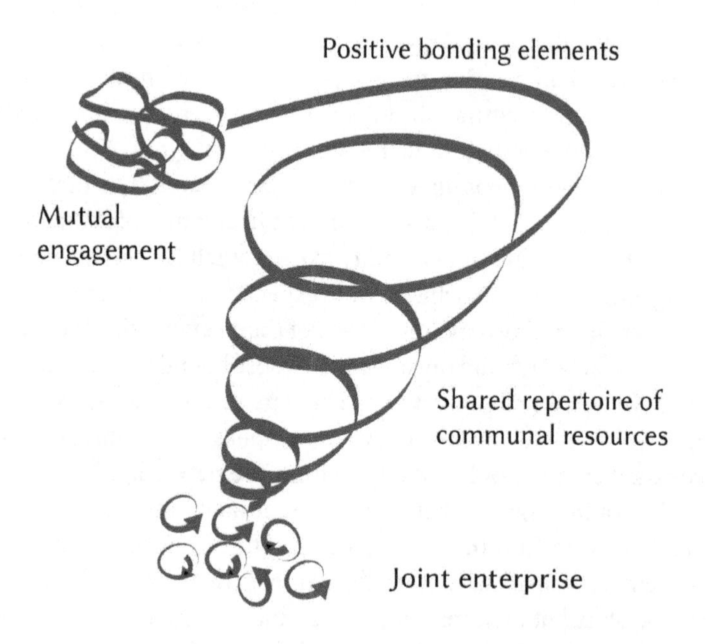

Figure V-1: Emergence of a community of practice

Of importance in relation to networks such as Sampat's, for instance, are the amplifying processes that those will embrace once they are established. Those processes are not to be chosen, they just are, yet they can be studied and facilitated. Once the network is established as a decentralized turbulence, it is understood that it will have to reach a further state of complexity to become a vortex, for its chaos to become organized (Figure V-1). Management theory refers to this as the emergence of a community of practice. Of importance here is the difference in complexity between a turbulence/network, and a vortex/community of practice. Both networks and communities of practice rely on amplifying elements, yet communities of practice emerge within more complexity than networks. Communities of practice can be defined as "self-generating social networks, referring to the common context of meaning."[217] As Etienne Wenger states:

> *"[m]embers of a community are informally bound by what they do together—from engaging in lunchtime discussions to solving difficult problems—and by what they have learned through their mutual engagement in these activities."*[218]

A community of practice defines itself alongside three key dimensions: a joint enterprise, which is to be renegotiated constantly by all members, a mutual engagement that binds all members together, and a "shared repertoire of communal resources that members have developed over time."[219] In terms of joint enterprise, this can be established by a shared vision, whose resulting actions will depend on the constant consensus of the community. The community of practice takes the lessons learned from past mutual involvement and acts on the basis of this shared knowledge. As within a network, flows of information and knowledge creation or sharing remain at the core, yet in a more complex manner. Information is not bare, in a constant state of re-generation, but takes into consideration the experience of the community's practice. It is expressed through collective memory, as a meme. Information relays do not stem from a direct cause and effect mechanism, but to a meaning in difference that connects to its joint enterprise. Capra explains: *"[m] eaningful disturbances will get the organization's attention and will trigger structural changes."*[220] This brings us back to the cybernetics introduced in Chapter IV. Information as a difference within a difference is understood as a meaningful impulse that will trigger a response. This meaningful impulse in the case of Sampat is someone's situation that warrants her community's assistance. In order to obtain a closer understanding of the elements that make and sustain a community of practice, let us look more closely at Anonymous and the amplifying elements within it.

Yin and Yang

Quoting management theorist Peter Senge, Capra argues that what marks the difference between a Cartesian mechanical organization and a living one, a complex adaptive system, are two key elements: a strong sense of community, and an openness to the outside world.[221] One day, I received an electronic message from "Me Ya" entitled: "You can haz VPN". It stated: "Part of safe surfing is having a VPN = which stands for 'Virtual Private Network'. When u [you] are on one very few ppl [people] can figure out who you are or where you are coming from. Some friends were kind enough to set one up for you." This became the set up for my first day of contact with the #OpTeach folks. While using this VPN, Anon people would drop by and answer my questions, join our conversations, leave, or go idle. Some would be more serious than others, as they told me how they enjoyed misleading journalists who ventured on their channels once in a while.

I virtually met with people from the Anonymous as well as the Telecomix network. While Anonymous is an encrypted network of people who are, well, nameless, Telecomix is an open network where members can participate using disclosing their identity. While the first can be seen as committing illegal acts, the other stays within legality. My long-term interlocutor from Telecomix, Peter Fein, explained how both networks are complementary in their actions, using the metaphor of the Yin and Yang of Chinese cosmology.[222]

Anonymous, the Yang, has brought websites down with Distributed Denial of Service Attacks, more commonly known as DDoS attacks or DDoSing, which involve overloading a server with requests to connect to a website in order to either significantly slow the server's response, or jam it. Telecomix, the Yin, is specialized in restoring services that have been arbitrarily cut off. During #OpEgypt in January-February 2011, as Anonymous was attacking the Egyptian government websites,

> *"the Mubarak Regime pulled the plug on the only connection in the country, and Telecomix was responsible for restoring Internet communications through a network of 1980s fax machines. We sent faxes to all the numbers we could locate in Egypt, explaining how to restore their connections so that they could organize their protests on social networks. We also sent them a demonstration kit with first aid instructions and Gene Sharp's techniques for non-violent action."*[223]

Both organizations, complementarily, have one clear common vision or motto: Internet freedom. Under this banner, people of all walks of life have joined to ensure that their common vision is respected. This is the only common denominator that exists between them. They have no leaders, no hierarchy, no funding, and no human face, yet they are occupying our virtual space and contribute to setting a certain political agenda.

Anonymous and Telecomix do not evolve in a traditional hierarchical system where their agenda is set by a few and executed by others. There are no members who are given privileged access over others, who receive rewards over others. They have no representatives, only interlocutors that are more amenable than others to talk to the press. The flow of energy in this network does not stem from its center. Instead, it is located within each Operation (#Op) that is undertaken by a group of like-minded people. Operations can be established in straightforward ways. If an Anon member finds an issue related to Internet freedom to be of interest, he or she will prepare a introductory package to raise awareness among others. At other times, several Anon members might discuss a common issue on an #Op channel, and may decide as a common front to open another #Op channel related to that issue.

One common inaccuracy made by observers of the network is to focus on a specific political message or set of issues. Anon is not specifically or directly a defender of Western-type democracies, human rights, or even the 99% of the Occupy Movement. As Anon3 points out,

> *"we are hackers, and we love to hack. We will do this no matter what, but as part of Anon, we get to defend what we believe in, Internet freedom. In sum, we do what we love to do, and we do something meaningful connected to it."*

Anon can thus be understood as a collective of people, primarily hackers, who congregate to hack for a cause they all subscribe to, Internet freedom. This cause is the icing on the cake of an overwhelming yearning to hack. The first Anon "operation," which took place in 2008, was against the Church of Scientology's forceful removal of a controversial video of US actor Tom Cruise, ranting about his and his Church's mission to *help* others, especially addicts, as a sacred duty to the world. Cruise advocated using any possible means, including "ruthless" ways, so that people can be saved from themselves.[224] This video, which was originally designed for internal recruitment but later leaked to the outside world, was tainted with a "you are either with us, or against us" mentality. The crazed belligerent tone of the actor was meant to impress prospective recruits on the verge of joining, not the average person. As it clearly manipulated right hemispheric thinking, it became an embarrassment to the Church of Scientology establishment as it plainly demonstrated the extreme position of some of its most high profile members. A clear demonstration of cult brainwashing, it had to be removed due to the bad PR this was bringing the organization.[225] Anon3, an ordinary family man who had been bothered by scientologists before, and knew of some people whose lives has been destroyed by the cult, decided to do something about it, and with some of his peers, created Project Chanology, more commonly known as Operation Chanology, a direct response to the Church's decision to sue the Internet hosts of the leaked Tom Cruise video. In Mid-January 2008, several websites of the Church of Scientology were subjected to trolling, with DDoS attacks, prank phone calls, and black faxes being sent to several Church facilities as a way to disturb their daily operations. According to Gabriela Coleman, this trolling consists of an

> *"unpredictable combination of the following: telephone pranking, having many unpaid pizzas sent to the target's home, DDoSing, and most especially, splattering personal information, preferably humiliating, all over the Internet."*[226]

Of importance to trolling is that it occurs *en masse*, in a focused way, over a very short time span, and is practiced – not designed – to not only overwhelm, but also paralyze its target. Following this first trolling phase, a video was then posted on the net, where Anonymous clearly established their goals and mission against the Church. While the Internet freedom motto is utilized in this video, it also lists more specific grievances against the church, and vows to expel the Church from the Internet altogether.[227] Later, a web campaign, as well as a website, were established as a way to inform the rest of the world about the scope of the operation, all this visibility gathered public support outside the Anon community and became instrumental to a series of non-violent demonstrations that took place from February 2008 onwards.[228]

The virtual moved into the realm of reality. On the legal front, Operation Chanology attempted to have the Internal Revenue Service investigate the tax-free status of the Church, and while this might not have been a success, it set the issue on the public agenda, and more importantly, allowed Anonymous to become more visible as a community of practice.[229] Other operations followed in the next two years, until Anonymous reached global visibility in its #OpPayback to defend Wikileaks toward the end of 2010.[230]

The Do-ocracy of the Hive

What makes Anonymous a community of practice? Let us take the example of #OpIran. As a few Anon met on different #Op channels, some specific and others more dedicated to openly discussing operations with AnonOps, AnonNet, or VoxAnon, they started exchanging information about Internet freedom in relation to demonstrations in various parts of the Middle East and Central Asia. This created or renewed interest in a wave of #Op channels related to countries, such as #OpEgypt, #OpIran, #OpPalestine, etc. (Figure V-2).[231]

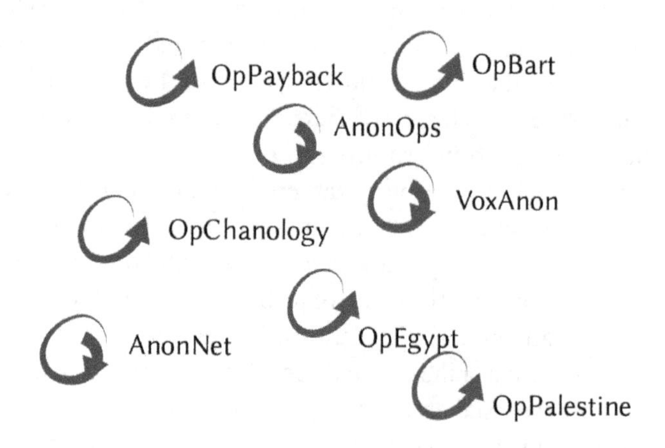

Figure V-2: Turbulence pattern of the Anonymous network

Of importance to the demography of those operations, according to Anon2, who introduced me to the community, is that the majority of Anons in those specific operations were women. This cannot be verified quantitatively, yet was reiterated on several occasions by various interlocutors. From this perspective, women from the region where the #Ops were taking place found the Anon community to be a safe space where they could assume a public, yet virtual, role. The Anon community, while originally composed of stereotypical hackers, i.e. young men spending day and night on the 4Chan network, soon became a more open community, primarily by way of online tutorials explaining how to take part in actions without necessarily having an Information Technology background.[232] Anon2 explains how #OpIran was created: "some guy showed up bitching about Iran [Anon1], got my interest, we opened a channel, trolled around for help, got a video made, put it on the net, it was a lot of work, there are ppl all over this net doing work like that, on all kinds of matters, not just this server, many servers." I then ask what decides on the popularity of some issues over others, and he answers that any issue can be amplified by a recurring visibility on a compilation of IRC channels, the web, and twitter.

The catalyst, or meaningful information in that sense, comes in the form of a video that is shared on the Internet between different members of the foreseen operation. Such videos are not necessarily proven in their origins or veracity, yet have a considerable power in appealing to one's right hemispheric sense of the whole. Anon1 shared a video with me that served as an information basis for #OpIran. It had been created during the post-electoral demonstrations of 2009, as the Iranian government was restricting Internet access in an attempt to diminish public support for the protests.[233] The video shows a street demonstration in Tehran being met by both the police force and men in black, understood to be the feared *nopos,* a governmentally appointed clandestine army geared toward the repression of the Iranian people. The video, shot from a rooftop, then shows the person filming it and an associate, trying to hide from a *nopo* raid on nearby apartment blocks, only to be allegedly arrested and, later, to die in prison. There is absolutely no corroboration of facts, evidence, or information in this video, yet its perception in relation to the predispositions of its audience toward the Iranian regime, has, for instance, recruited Anon2 to #OpIran.

Of importance here is the trust that members of the community have with one another in relation to the material that they post online. This relates to the mutual engagement of Anon as a community of practice. This trust, however, is not unlimited or blind. In the private messaging conversations that I have had with Anon1 and Anon2, it is clear that neither trusts the other beyond the common goal of the operation, yet they can easily relate to one another and share very personal stories. This trust dimension is rather complex, and I assume, pertains to the virtual nature of the community. While Anon1 and 2 have the bond that soldiers might have in a common operation, it is very different from the unconditional trust that I found between the members of the Gulabi Gang that I met. In relation to videos, over time, others are shared or placed on the web, and serve as a constant reminder to the Op members of the necessity of the operation. This relates to Wenger's joint enterprise attribute of a community of practice, which always has to be renegotiated between members. As an amplification mechanism, members place descriptions of the operations on the web, with supporting videos, frequently asked questions sections, detailed descriptions of the forthcoming operations, etc.[234]

As stated earlier, since it stemmed from a network, Anonymous as a community of practice does not evolve with a traditional leadership but functions as what Fein refers to as a adhocracy, meaning that the individuals who dedicate the most time and effort to the operation are the ones who, invariably, are going to steer the hive of Anons in a specific direction.[235] From this perspective, decision-making is not specifically consensus-based, as in the Occupy movement, but is a kind of hybrid between consensus and hive-like activity. Leadership relates to a swarm form of collective intelligence, the same that enables Leafcutter ants to eat my garden away.[236] Sometimes, members of the operation will not agree on a particular point, and this will lead to either one member leaving the group, a long argument with much trolling, or one part of the hive resolutely moving in a different direction, hence loosing the interest and support of another part of the hive. When this happens, in rare cases, this can lead to the auto-destruction of an operation. Parts of the hive can be directly attacked by another part; they will be, as Fein explains, "thrown under a bus." This actually occurred to #OpIran in April/May 2012. Fein recalls: "some felt that #OpIran was an offset of the Green Movement of Iranian dissident Mir Hossein Mosavi, yet we do not do politics, or regime change, we are strictly helping in Internet Freedom matters." The line is undoubtedly fine, and sometimes blurred, between an #Op for Internet freedom and political grounds. When this becomes too obvious, the node becomes a threat to the rest of the community, and self-regulates into oblivion. Fein explains that a "throwing under the bus" can be directed at an individual or a whole #Op; it involves

> *"some direct trolling, but also doxxing [digging up personal info] and telling their friends/allies the stupid stuff they've been up to [determining where this crosses to just spreading rumors is left as an exercise for the reader]. Throwing whole ops under is a bit different [...] I've seen that be as simple as persuading people to abandon an op, a channel takeover, extensive doxxing + coordinated public shaming/discrediting/ disavowal [via Twitter], or appealing to IRC opers [who admin the servers] to shut an op down. Dramatically throwing big ops under is rare more often, bad ops fail to launch to critical mass and just fizzle out. Bus throwing is expensive and not much fun for anyone."*[237]

Since virtual communities of practice do not involve physical structures, the auto-renewal of their system is much easier and less dramatic than, let's say, "purging" a particular dysfunctional node of the Gulabi Gang, or a renegade member of the Hezbollah. An Anon thrown under a bus may not see their physical life or environment dramatically affected, while a dysfunctional operation that looses support, i.e. the relationship that exists between members, will be able to morph into another #Op. From the perspective of panarchy, "throwing under a bus" can account to a release mechanism of the system. As a self-regulating collection of hives, the community of practice that is Anon enjoys a heightened resilience that allows it to move to higher stages of complexity, even moving beyond the virtual sphere. In that sense, the visibility of Anonymous as connected to the Occupy movement could add complexity to the community of practice.

Obsolescence

In the last few years, the emergence of social dissent has become more visible across the world. The main success of Wikileaks as a network has been to uncover many public secrets and to enable public discourse to move beyond the realm of conspiracy theories in terms of international relations and Western foreign policies. It has made my work much easier, uncovering the many double binds of the peace industry, as seen for example in the case of the Afghan *Bacha Bazi* discussed in Chapter I. Its main weakness, however, has been the swelling ego of one of its founders, Julian Assange, whose urge to appear in the limelight has dealt a significant blow to the organization.[238] Once any leaderless network allows for one of its members to assume its identity, it leaves the sphere of chaos and enters the Cartesian realm of traditional methods, where it can be easily contained and discredited. Think of it as a soccer match, where teams find it an advantage to play on their own turf. Since all human beings have their flaws, and in the case of Assange it is his male ego, the state found a way to "counter-troll" the network in ways that it knows best, the construction of a sex-tainted legal suit against Assange, the fastest way to initiate a massive loss of public support, discredit the network, diminish its financial resources, arrest and subject leak Bradley Manning to, according to a UN rapporteur, "cruel and inhuman" treatment, etc.[239] Since the Assange "scandal" erupted, the attention and information flows of the network have been disrupted, and, more importantly in relation to networking, the relationships between members have been altered. Seeing the manner in which Manning has been treated, who would entrust any information to a network on the brink of collapse due to the visibilisation of its self-appointed leader? From that perspective, Assange's decision to come to light was a tactical mistake. Still, as I mentioned earlier, Wikileaks has had an undeniable impact in exposing the many double binds of international relations. It has opened a window in popular culture for political dissent to be backed up by the ugly reality of Realpolitik. It has also brought to light the power that networking individuals may have in relation to the state. In the same way that the Arab spring created a bifurcation point for random individuals to realize that they can trigger social change, Wikileaks is an example of how open flows of information may be considered as a threat to any state worldwide, and more importantly, the state as a paradigm.

There is another association that should not be made in academic publications, as far as my Anonymous contacts are concerned: "terrorism." As soon as I entered in conversations on #OpTeach and #OpLearn, we began to discuss the issue of how the US media and government portray the organization in terms that are borrowed from the realm of insurgency and terrorism studies. In February 2012, the director of the National Security Agency, General Keith Alexander, warned that Anonymous *"could have the ability within the next year or two to bring about a limited power outage through a cyberattack."*[240] While this type of "terrorism" framing is no stranger to governmental rhetoric and scare mongering – recall US Secretary of state Colin Powell's ill-fated speech delivered at the UN to justify an impending invasion of Iraq – it is clearly intended to scatter the general public support for Anonymous, as well as scare some of its actual or potential members off. What makes Anonymous or Wikileaks so threatening to power structures? It is less their technical capacity than the potential for the transparency they advocate to become a public norm, as well as the potential for ordinary individuals to realize that they can exist and thrive as a community outside a conventional state structure. The same occurred as a result of the Arab Spring. Why didn't the Obama administration acknowledge or support the Egyptian demonstrations as soon as they occurred? Because the double bind that was exercised on Egypt by Barack Obama, who delivered a tear-jerking, hopeful speech on democracy while keeping the Mubarak dictatorship in power as a safe Israeli ally in the region, was bound to recede, leaving the state paradigm of both the US and the Egyptian governments powerless to control mass uprisings for freedom. The freedom that was asserted there, and supported by both Anonymous and Telecomix, was not only in relation to President Mubarak, it was entering Rumi's field beyond the state Paradigm of right and wrong. This represents the real threat of Anonymous and its avatars. In the same way that al-Qaeda once operated as a center-free turbulence, before being transposed back to a hierarchical/Cartesian mode by the state paradigm, Anonymous represents the potential for political and social activism to exist outside the state mirage of "democracy."

Fein has been delivering speeches on the adhocracy, and how the practices of the Anonymous and Telecomix community have started to show that in the present, the right hemisphere, change is occurring, immanent, and unavoidable.[241] He explains that adhocracy

> *"exists in the spaces left over and in between, [the invisible beyond the map]. It's a politics of practice and oral tradition."*

From this immanent perspective of the community of practice, the state Paradigm of Cartesian control becomes obsolete. This is the real threat. I believe, therefore, that the NATO intervention in Libya, to "protect" civilians (apart from migrant workers, women, Gaddafi supporting tribes, etc.), was a state Paradigm attempt to re-capture the people-power social change movement and place it on its own "turf," that of the left hemisphere. The organization of elections in Egypt has had the exact same function, since the democracy meme belongs to the future, not the present. The map of one man, one vote, and the illusion of the present is never the territory of the government formed after an election, which only reproduces more of the same and benefits the same elites.

Systems of Influence: When the Anomaly Becomes the Epistemology

When does a community of practice reach another bifurcation point that will enhance its complexity, and allow it to reach a renewed yet deeper state? The recent visibilisation of adhocracies, or communities of practice, which initiated in the Arab Spring and culminated in the West with the Occupy movement, could yet become an even greater threat for the state Paradigm (Figure V-3).

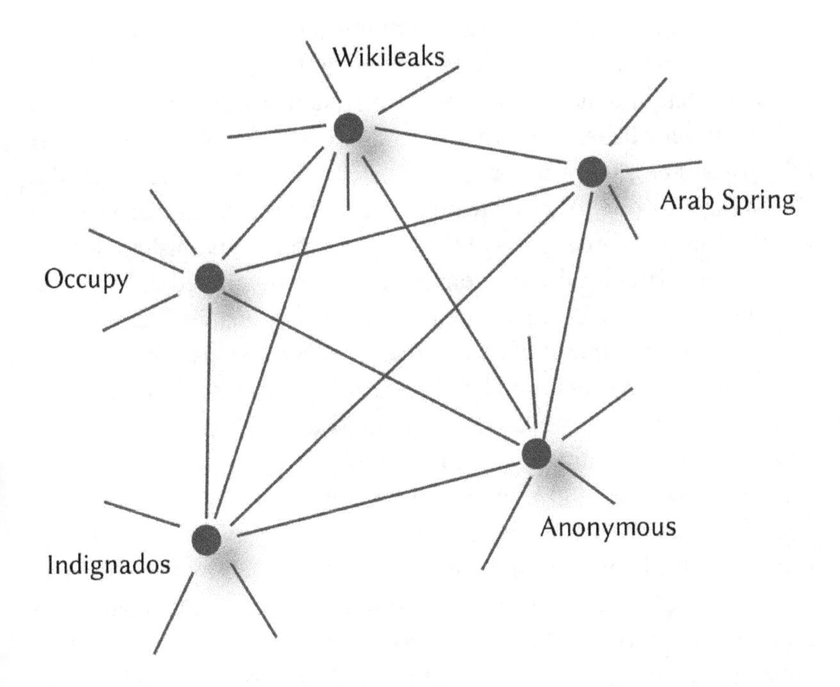

Figure V-3: Communities of practice for "freedom"

Margaret Whitley writes about the bifurcation of communities of practice into systems of influence, and defines it as the third phase of emergence, occurring when "[t]he practices developed by courageous communities become the accepted standard." She explains:

> *"[p]eople no longer hesitate about adopting these approaches and methods and they learn them easily. Policy and funding debates now include the perspectives and experiences of these pioneers. They become leaders in the*

> *field and are acknowledged as the wisdom keepers for their particular issue. And critics who said it could never be done suddenly become chief supporters (often saying they knew it all along.)"*[242]

When adhocracy becomes a way of life, a standard, when ordinary individuals no longer surrender their freedom through "democratic" elections to a state and the elite it strives to maintain, is when adhocracy and people-power will become a system of influence. The visibility of complex adaptive systems for peace or social change may well set public agendas for change, through the witnessing of real, positive, and manageable alternatives from other parts of the world. A conscious epistemological shift, uncovering local-local consensual initiatives based on different parameters that are deemed "useless" to conventional indicators, such as the bonding elements of care, esthetics, relationships, etc., may well set an agenda for experiential change, a step-by-step evolution.

While decolonizing peace, as a community of practice, has not reached a systemic level of influence, communities as complex adaptive systems must be encouraged to remain as they are, in the right hemisphere, and not sell out to the left hemispheric Cartesian way of thinking. This awareness is already present, yet always threatened by the efforts of the older, incomplete, Cartesian paradigm, to reclaim them back into its symmetrical reality. From this perspective, the Cartesian paradigm behaves like the mainstream perception of a black hole – which is proven to be more complex than this –, eating the energy present on its path, hence the valid uneasiness that many have in relation to bringing an alternative reality into the realm of the visible.

In relation to the Arab Spring, one observer commented emphatically on how he had supported political change in the region all along: Usama bin Laden. In his Letters from Abbottabad, he states:

> *"[t]hese events are the most important events that the nation has witnessed for centuries, (...) so if we double the efforts to direct and educate the Muslim peoples and warn*

them from the half solutions, (...) the oncoming stage will be for Islam."[243]

Originally a turbulence that never became a vortex, al-Qaeda lost its influence in the same way as the state Paradigm (albeit temporarily): through the collective awareness for change that was prompted by the Arab Spring. As the state Paradigm's primary symmetrical adversary, global insurgency, al-Qaeda was as threatened with obsolescence as the state was, since the presence of local-local communities of practice was no longer part of the traditional equation of security and liberal peace. As random individuals realized that they were able to assert themselves for social change, they were also able to appreciate that they no longer needed others to represent them. The electoral meme won those communities back, as a strong remember connection from the higher level of panarchy, and through this, so did a certain vision of Islam, with the early claims of electoral victory from both the Muslim Brotherhood and the former regime.[244] Only time will tell how the Tahrir Square community of practice will survive amidst those state/epistemological recuperations. The meetings that recently took place in Jordan between the Egyptian Muslim Brotherhood and al-Qaeda in Iraq, supported by the Turkish government, certainly quell the early "we did it" euphoria prompted by a simplistic reading of the Arab Spring.[245] Still, is the emergence of adhocracy and dissent as a system of influence possible? Will transparency and adhocracy as an anomaly ever become a paradigm?

The answer to this question may well lie in the importance of dynamic processes as opposed to theoretical ends. Amplification is not an end; it is a process that can only occur through panarchial cycles of renewal. Adhocracy or transparency as theories cannot become a paradigm. They would risk becoming instantly fossilized, since theories in social sciences are by definition dead ideas. Adhocracies as pathways, practices, and panarchy cycles may become a system of influence, rendering the state Paradigm obsolete in its exclusiveness. Adhocracy may be the complexified territory of the state map. Does this mean that states in general cannot coexist with adhocracy, in the same way that the UN cannot coexist alongside decolonizing or hybrid forms of peace?

Panarchy reminds us that there is a need for a remember connection, and so does Fein:

> *"[w]hen you have an obsolete legacy system on the edge of collapse, it can be tempting to just throw it out and start fresh. But experience with technology teaches us to avoid this if we can – revolution is expensive and risky. It doesn't work so well when you have a large, dependent user base and it often turns out you know less than you thought you did going in. Instead, you have to replace a legacy piece by piece from the bottom up. We need to starve the beast – not of dollars, but of our personal investment and energy. It took a long, long time to get into this mess, and the struggle to rebuild a better world is going to take the rest of our lives."*[246]

An alternative epistemology for peace and conflict studies does not lay in the establishment of yet another theory, but in the realization that a different view of the world and our surroundings, a paradigm shift, allows us to see a different reality that already exists and can co-exist with an awareness of the "help" addiction of the remember functions of the higher panarchy levels. From this perspective, how can a different worldview be safely brought into the realm of the visible?

Dialogue as an Amplifier

How many of us have taken part in workshops, conferences, or meetings where the same key words and categories of knowledge were brought forward, debated, slightly challenged, and at times enhanced? We find originality in one concept pertaining to development to be re-used in another sphere; let's say, media studies, when we rephrase an old concept or add new levels to our categorizations of knowledge. Where is the originality in this? The current state of knowledge in the field of peace and conflict studies pertains to the categories that form it. When we reach the "field", or engage with a liberal-peace-trusted interlocutor, we expect our language, maybe within variations, to be spoken. In fact, we will find our interlocutors and grant them our trust in form of expertise recognition only if they know our language, or have attended our academic formations. I have often met with colleagues from the South, only to have them immediately expose their credentials from Bradford, George Mason, Uppsala, and, of course, UPEACE, hence being "kosher," so to speak. How open are we to engage in a deep exchange with people whose language we neither recognize, nor consider as being valid, outside our own?

It has not escaped me that in ascribing the names or categorizations of chaos, panarchy, or cybernetics to some processes, some may also fall into that trap. Yet replacing a panoply of keywords by others is not a paradigm shift, it is merely a re-situation of knowledge. The proposed paradigm shift here is the use of a different type of lens through which to visibilize the invisible. Dialogue can be a process toward that state.

A dialogue is not a discussion, nor is it a conversation.[247] While engaged in a discussion, interlocutors will primarily pit their arguments against those of the other. They will spend most of their time thinking about their next sentence, without listening to the contributions of their fellow discussants. A discussion is a form of debate. It pertains to Bateson's realm of the symmetrical. It generates a tit-for-tat exchange that will seldom bring participants to a consensus. Voices may be raised, tempers may be altered, and grudges may be held long after the discussion has taken place. A conversation is a less controversial form of discussion, still symmetrical, but harboring a less opinionated style. Both discussions and conversations address a specific topic that has been established beforehand and whose parameters are rarely challenged. There is no power in either form of interaction, since their agenda has already been set. In relation to Lukes' view of power, both decision-making and influence are less important than agenda setting.[248] After all, what matters about a decision if it does not exist in the first place? Dialogue comes as an alternative to conversations and discussions. While conversations and discussions can be understood to be an exchange of information, a dialogue revolves around the joint creation of new knowledge. Bohm defines it as creating something new, through the establishment of a common space within which participants, having suspended their assumptions, will be able to share and create at the same time. For a dialogue to be able to take place, no agenda must have been set in advance.[249] Participants will be able to meet with no set of assumptions on a particular topic, which in itself will allow for a exchange to be able to take place in the form of collective consensual agenda-setting. Of interest in the processes of both Anonymous and Telecomix is the amount of time spent chatting on various matters completely unrelated to the operation supposed to be taking place. This allows for a community agenda to be established, and for Lukes' third dimension of power to be asserted. This communication format also allows for the network's flow of information and creation of knowledge to maintain it as a live organism.

Dialogue is a vital process in relation to decolonizing peace, yet it also represents an epistemological shift in the manner in which research, social encounters, information sharing, and writing occur on a daily basis. Time, budgets, logistics, and "expertise" will often be blamed for the impossibility of a dialogue process to occur; yet dialogue can very well be employed as a pathway toward the establishment of a mixed symmetrical and complementary relationship within decolonizing peace. Imagining dialogue within a conference format, moving beyond the sphere of open space technology that still frames a conversation within a given agenda, is also a future expression of decolonizing peace at an academic level. When reaching the "field", this would entail a different format of consultations between donors and communities, "experts" and local elites, etc. Dialogue thus becomes an amplifying process of complex adaptive systems, crucial for decolonizing peace. The shift from strategic planning to scenario planning is also a form of dialogue, based on bringing the invisible to light by devising possible future scenarios and preparing accordingly. Dialogue, in decolonizing peace as a process, can be utilized to infinity, i.e. forum theatre, folkloric singing, etc. Dialogue as an amplifying process also relies in a holographic manner on the bonding elements unpacked in Chapter IV. A book is currently being prepared on the ethics and esthetics of hacking in the Anon community that will no doubt provide an in-depth complement to the present chapter, as well as a recursive loop into Chapter IV's dimension of the "sacred" kinesthetic understandings of decolonizing peace.[250]

Conclusion

Given the recent developments of Cartesian power structures reclaiming social change onto their own territories, elements of awareness also need to be amplified in relation to decolonizing aspects of peace. It takes more than a leap of faith to privilege a decolonizing practice of peace. Within our Cartesian paradigm, faith has been placed in the sphere of the irrational. Some of the students that go through a whole course of decolonizing peace recurrently bring the same cliché back to our collective agenda: "what about the real world?" This book does not assert that another world is possible, but that it is already there, everywhere, and that should we chose to ignore it, it will continue to thrive without us. At the risk if sounding like an illustrious, bearded revolutionary from Germany, what the spillover of the Arab Spring into the Occupy Movement ought to have brought us is the realization that the 99% in the greater South need not be marginalized by the 99% in the greater North.

The possible bifurcation of communities of practice into systems of influence depends on awareness, faith, and also a conscious choice toward a collective evolution for change; adhocracy can also permeate academia (Figure V-4).

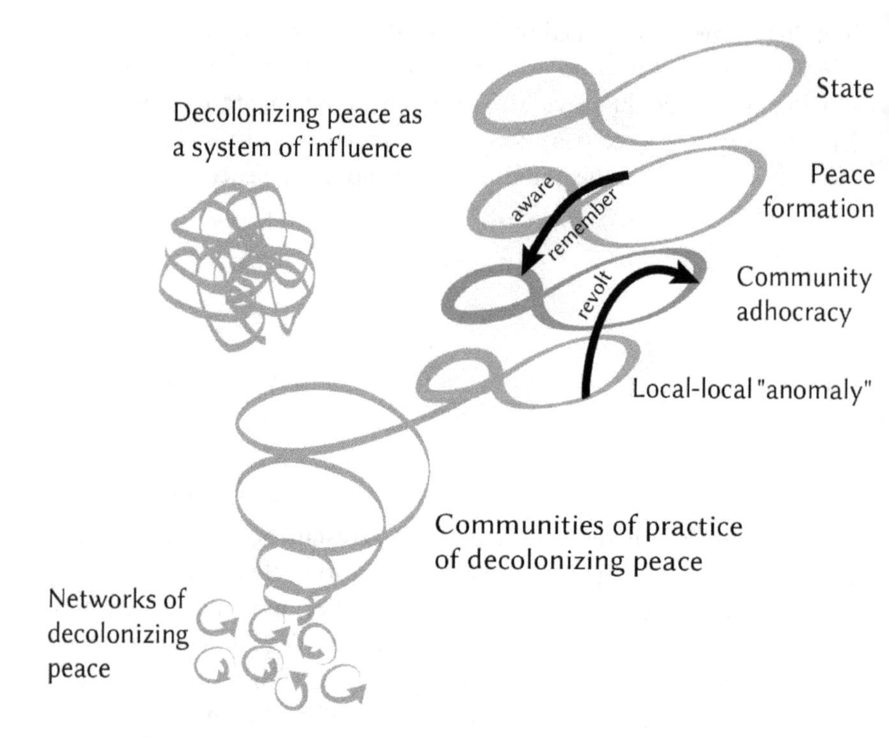

Figure V-4: Decolonizing peace in three stages of emergence

This book has not sought to introduce an alternative epistemology to a community that is not ready for it, yet the real challenge will be the bridging between a right hemispheric vision of peace, and the rest of the peace community, industry, etc. Awareness as well as a conscious resort to dialogue as a practice of collectively creating something new seems to be a necessary step for the discipline of peace and conflict studies to be able to positively act as a remember connection in relation to the local-local practice of decolonizing peace. Change is already occurring outside of academic circles. It can be facilitated through the practice of a "sacred" connection.

Epilogue

Paola, Brandy, Brittany, and Justin are between three and eight years old. Their parents, from Nicaragua, are illegal laborers in a coffee plantation that sells its harvests to the US-based Starbucks coffee company. They love school, playing together, and occasionally swimming in the nearby river. Their main concerns in life are to be able to attend school, to find their toys untouched when they return home (since their Costa Rican neighbors regularly visit their homes to steal anything they find useful), and walk to the latrines without being bitten by the neighbor's dogs, which regularly attack them. Their house does not have running water or electricity and amounts to a few wooden boards put together, making it look more like a barn than anything else. When their parents do not have enough money to buy diesel for their generator, the water pump that connects their house to the river does not function, and this means they need to walk a long distance to access a source of water. Since many Costa Rican rivers are polluted, especially those adjacent to coffee plantations, they often fall sick with stomach parasites. Still, they consider this routine to be the best part of their year.

When the coffee-harvesting season comes, for about four months in the year, their lives change drastically. They are expected to help their parents pick coffee for several hours a day, and have to surrender their living area to another 100 migrant workers. This means packing their toys away and sharing a room with an of average 12 other workers who are unknown to them. The work they are forced to do is tedious, tiring, and they are often afraid of being bitten by snakes. When they return home to rest, they play together for a while and then crash on the wooden boards that they are supposed to sleep on, too tired to even be uncomfortable.

For the sixth year in a row, Starbucks has been named by Ethisphere as one of the "world's most ethical companies."[251] Starbucks' website boasts of a strong corporate responsibility policy, recognized business ethics, and compliance with international standards and compliance with international standards.[252] Since the Costa Rica legal system prohibits children from working, this is taken at face value by Starbucks, which, according to the workers, have never been seen visiting the plantation. Should I bother to write, once again, that "the map is not the territory?" One can understand that a plantation based in Costa Rica might be located rather far geographically, hence Starbuck's convenient ignorance translating into award-winning corporate bliss. However, how about the plantation's proximity to my own academic institution, UPEACE?

Paola, Brandy, Brittany, and Justin live five minutes from the University, yet it took seven years for any student to take interest in what happens next door to them.[253] Our students take buses in and out of the university every day, pass by the plantation and see the workers, adults and children, lining up near the road four months in the year, and yet they choose to look the other way, since websites such as "Ethisphere" tell them that all is fine and quiet on the Southern front. We boast of our Responsible Management program, we benevolently give English lessons to the Costa Rican children of a nearby state school, yet we do not take the time to stop and smell the poisonous flowers of our tropical countryside. We feel uneasy, powerless, about the enormity of the human rights violations that occur all around us. In the past, when asked by our faculty about their lack of interest for the coffee plantation workers, our students gave us many reasons for their non-involvement, the most infuriating of all being the need to maintain a good relationship with the government of Costa Rica. As if our host government is solely responsible for the lack of ethics of local businesses. Peace and conflict studies generally focus on conveniently labeled dysfunctional societies as a prime territory to be "helped," often making abstractions of what happens on its doorstep. After all, it is easier to talk about atavistic ethnic hatred in Rwanda than look into what happens next door.

Yes, there are ethical considerations to take into account regarding how to approach such a sensitive issue without harming the families in question, in evaluating the social and economic consequences of a potential research into their working conditions or legal status, etc. Yet, the families in question have welcomed the relationship that they have now built with our students, not the "help" that we could have easily given them, in discarding our old clothes to them, etc., but the visibility and networking potential that our growing relationship has initiated. They no longer feel alienated as illegal migrant workers in Costa Rica; they feel part of an unfolding community. There is a long way to go for our relationship to transform our daily realities, and for the children to be able to grow up in a safe environment. Yet a dialogue has been opened, and the possibility of creating something new together is part of our present reality. The steadfast nurturing of a community relationship that moves beyond the pre-established liberal peace parameters of contributing to our environment from the top down has taught us that any open collective involvement needs time, trust, care, etc. In fact, the relationship that we are building together is so far focused on the bonding elements of complex adaptive systems. The visibilisation that raises emotions on both parts may be one of the first steps toward the formation of a turbulence for change.

This book has sought to provide an alternative epistemology for peace and conflict studies. It originates from the dialogical observation of the cruel effects of the double binds of the peace industry in terms of the resurgence of human trafficking, sexual abuse, modern slavery, etc. This observation was initiated in spite of me, and solidified in a research project over many years. It is in initiating a dialogue with key actors in different complex adaptive systems for peace that decolonizing peace has taken its present, transient shape. Decolonizing peace, through an epistemological shift, can be understood as a transitory concept that highlights the invisible aspects, processes, and practices of peace worldwide, but especially in the global South. Decolonizing peace places an emphasis on all the key elements that do not constitute any typical categories of the liberal peace paradigm. Those categories, "useless" since they cannot be quantified or made part of a conventional budget line, can be seen through the lens of decolonizing peace as the bonding, amplifying, and stabilizing elements of a complex adaptive system, whether they have taken the form of a turbulence/network, a vortex/community of practice, or a panarchy/system of influence. These categories, unable to be quantified or named from a conservative liberal peace perspective, rely on the "useless," and amount to dialogical expressions of love, care, esthetics, ethics, emotions, inspiration, inclusion, etc. From this perspective, the "useless" makes matter, since it cements relationships, nourishes flows of information and the elaboration of new knowledge that keeps the pattern of relationships intact, within a structure of constant renewal. The "useless" becomes the "useful," from the perspective of decolonizing peace.[254]

It is undeniable that liberal peace currently represents a system of influence, a paradigm that shapes responses to disasters, foreign policies, humanitarian assistance, and the apprenticeship of peace at an academic level. This paradigm has been increasingly challenged in the last few years with the naming of it as such: liberal peace, the establishment of an anomaly level within its panarchy, the liberal peace dialogue initiated among scholars and practitioners, and the current establishment of another anomaly level in the peace panarchy in terms of Richmond's concept of peace formation.[255] For now, decolonizing peace represents yet another anomaly level forming within the peace panarchy. It brings visibility to local-local initiatives for peace that evolve with the traditionally unquantifiable sphere of the "useless." It also calls for the conscious establishment of an awareness of the Northern addiction to "help," which invariably evolves within the expression of a pathological double bind in the Northern relationship with the rest of the world. While decolonizing peace remains aware of the necessity of both revolt and, more importantly, remember connections in the peace panarchy, it contends that no remember connection can be asserted outside a double bind if there is no awareness of the "help" addiction. From this perspective, the remember connection no longer translates into "help," but an openness to dialogue, from the perspective of the creation of something new. Decolonizing peace connects the visible and the invisible within the community of practice, while it can also raise awareness within the panarchy/system of influence of peace, thus facilitating the subsequent emergence of the liberal peace panarchy or system of influence into a post-liberal or peace formation system of influence. It is most likely that this emerging system of influence will be consolidated within the realm of post-liberal peace, as the state system will find it difficult to relinquish its power. More importantly, it remains vital that this emergence is not perceived by the state Paradigm as a threat to its existence. This would imply a risk of suppression by the system based on a perceived assault on its resilience capacities.

Such an assault might not be an open confrontation, but may well exist in terms of more "help." Take the resilient capacities of the Gulabi Gang as existing in the formation and nurturing of an open circle of help. Should an international body offer significant financial assistance to the community, or a political mandate, this could have a disastrous impact on the foundations of the system itself. It is in this sense that the presence of hybrid actors on the ground is crucial, in mitigating the relationship between liberal peace actors and local-local complex adaptive systems based on the "useless." An alternative epistemology of peace can bring hybridity forward as a strong link between the revolt and remember connections of the peace panarchy, whether it be liberal or post-liberal.

Everything remains to be achieved in terms of decolonizing peace and the emergence of a post-liberal peace system of influence. A strong ethnography of the interface between hybrid local actors and the liberal peace system must be carried out, an ethnography of the dialogical relationship between local-local complex adaptive systems as the Gulabi Gang, Samawada, even Anonymous, must be carried out from the epistemological lens of decolonizing peace. The research tools that are currently presented to many peace and conflict graduates across the world are based on a symmetrical Cartesian understanding of the world, whose situation of knowledge from a Northern standpoint remains full of condescension, judgment, and arrogance. From this perspective, alternative research methods for peace and conflict studies must be established, alongside the already established Decolonizing Research Methods, which focus on dialogue in relation to the re-formation of categories of knowledge.[256]

Resistance to an alternative epistemology not only exists in the global North, but also in the Southern elites educated in the North, anxious about being made redundant by the importance given to the local-local complex adaptive systems for peace. This is, of course, a misperception of the potential that exists in decolonizing peace, as who else but Southern-born and -based academics are best qualified to carry out ethnographies of local-local complex adaptive systems?

A colleague of mine recently pointed out that many academic links between Africa and Central/Latin America are channeled through the global North, in the same way that air transport connects both regions of the world through Northern hubs. Peace and conflict studies no longer belongs to a Northern epistemological hub, since local-local complex adaptive systems are providing an unavoidable alternative reality to the Northern map. It is time for Southern-based academics, myself included, to stop looking North for direction and encouragement. We have what it takes to re-shape our discipline and be more than mere illustrations of Northern-established concepts. The resistance shown by colleagues who would rather go North than write their PhDs in the South belongs to colonial times. Those days are long over, except in our minds.

In April 2009, I was asked to teach a course on insurgencies and terrorism at Benhares Hindu University in Varanasi, India. As many Kashmiri colleagues had complained to me that their realities seemed to be invisible to the rest of India, I chose to bring only one book for my entire class to read: *Curfewed Night* by Basharat Peer.[257] This book narrates the childhood of a boy in the occupied part of Indian Kashmir. While it depicts the perceived daily humiliation that brings some to take sides in against the Indian state, the anomalies, it also provides a compassionate account of the positions of other actors to the conflict, such as the Hindu Kashmiri pundits. As I sent my book request to colleagues in Varanasi, I was immediately cautioned by a junior member of the faculty that some students in the classroom might report me to the Bharatiya Janata Party, a Hindu nationalist extreme right party. I did not see what the problem was, and chose to make a non-issue out of this possibility. The delivery of my course came without major problems, so I chose to add possible insult to injury in presenting the Gulabi Gang to both my class and some members of the Gender Studies department.

While the Gulabi Gang, by then internationally recognized, was operating only a few hundred kilometers away from my classroom, and by Indian standards, this is right next-door, no one had ever heard of them, not even the Gender Studies department, too busy translating the works of Simone de Beauvoir to study its surroundings. As I presented Sampat, one of my students stated that their peace and conflict studies curriculum needed to be decolonized and brought closer to home. Most of their theoretical input came from Europe and the US. The only Indian token in their program, until then, had been Mahatma Gandhi, and they were not all unanimous about his "holiness" either. One of them stated that while it was time that their own reality was being taught to them, they should also, as graduate students, become ready to step out of their classrooms to see what was happening around them, not through western eyes, including mine, but their own. As they highlighted the need for a decolonization of their education, I stressed the need for a decolonization of our own minds as academics. The invisible only remains as such within our compliant minds.

Decolonizing peace is not a utopia, but a reality; it is in the now. As scholars, stepping into our right hemisphere, and linking the right and left hemispheres through "the sacred", can enable us to re-shape our discipline as an inclusive North-South holistic endeavor, Northern awareness of its "help" addiction and Southern appropriation of a long overdue intellectual equity. We might not overcome the state Paradigm, but in the same way that Sampat chose to act in relation to immediate circumstances, the double binds we can no longer ignore compel us to decolonize our minds as students, scholars, and practitioners.

Bibliography

Achebe, C. (1994). Things Fall Apart. London, Anchor Books.

Alagha, J. E. (2006). The Shifts in Hizbullah's Ideology: Religious Ideology, Political Ideology, and Political Program. Amsterdam, Amsterdam University Press.

Alagha, J. E. (2010). "Wilayat al-Faqih and Hizbullah's Relations with Iran." Journal of Arabic and Islamic Studies **10**: 24-44.

Alagha, J. E. (2011). Hisbullah's Identity Construction. Amsterdam, Amsterdam University Press.

Alagha, J. E. (2011). Hizbullah's Documents: From the 1985 Open Letter to the 2009 Manifesto. Amsterdam, Amsterdam University Press.

Allah, H. (1985). Nass al-risala al-maftuha allati wajjaha hizb allah ila al-mustad 'afin fi lubnan wa al-'alam. Amal and the Shi'a: Struggle for the South of Lebanon. A. R. Norton. Austin, University of Texas press.

Anderson, M. B. (1999). Do No Harm: How Aid can Support Peace -Or War. Boulder, Lynne Rienner Publishers.

Azani, E. (2008). Hezbollah: The Story of the Party of God. London, Palgrave.

Bales, A. and R. Soodalter (2009). The Slave Next Door: human trafficking and slavery in America today. Berkeley, University of California Press.

Barnard, A. (2011) "Occupy Wall Street Meets Tahrir Square." New York Times.

Bateson, G. (1972). The Cybernetics of "Self": A Theory of Alcoholism. Steps to an Ecology of the Mind. G. Bateson. Chicago, University of Chicago Press.

Bateson, G. (1972). From Versailles to Cybernetics. Steps to an Ecology of the Mind. G. Bateson. Chicago, The University of Chicago Press.

Bateson, G. (1991). Ecology of the Mind: The Sacred. A Sacred Unity: Further Steps to an Ecology of the Mind. R. E. Donaldson. San Francisco, Cornelia & Michael Bessie Book.

Bateson, G. (2000). Form, Substance, and Difference. Steps to an Ecology of the Mind. G. Bateson. Chicago, University of Chicago Press.

Bauman, Z. (2001). Modernity and the Holocaust. Ithaca, Cornell University Press.

Blanford, N. (2011). Warriors of God: Inside Hezbollah's Thirty-Year Struggle against Israel. New York, Random House.

Boff, L. (2008). Essential Care: An Ethics of Human Nature. Waco, TX, Baylor University Press.

Boff, L. and M. Hathaway (2009). The Tao of Liberation, Exploring the Ecology of Transformation. Maryknoll, Orbis Books.

Bohm, D. (1996). On Dialogue. London, Routledge.

Bolkovac, K. and C. Lynn (2011). The Whistleblower: Sex Trafficking, Military Contractors and One Woman's Fight for Justice. London, Palgrave Macmillan.

Bolte Taylor, J. (2006). My Stroke of Insight. New York, Plume.

Bose, S. (2003). Kashmir: Roots od Conflict, Paths to Peace. Cambridge, Harvard University Press.

Bowyer Bell, J. (1996). Terror out of Zion: The Fight for Israeli Independence. New Brunswick, Transaction Publishers.

Brandstetter, R. H. and V. C. Fontan (2005). Final Report for Political Process Assistance Review. Monitoring and Evaluation Performance Program, Phase II (MEPP II). I. International Business & Technical Consultants. Washington, D.C., United states Agency for International Development.

Briganti (de), G. (2012) "The Real Reasons for Rafale's Indian Victory." defense-aerospace.com.

Briggs, J. and F. D. Peat (2000). Seven Life Lessons of Chaos: Spiritual Wisdom from the Science of Change. New York, Harper Perenial.

Bulhan, H. A. (2008). The Politics of Cain: One Hundred Years of Crises in Somali Politics and Society. Bethesda, Tayosan International Publishing.

Caplan, G. (2008). The Betrayal of Africa. Toronto, Groundwood.

Caplan, G. (2009). Obama and Africa – a major disappointment. The Nation. New York.

Caplan, G. (2012) "Peacekeepers gone wild: How much more abuse will the UN ignore in Congo?" The Globe and Mail.

Cappelli, R. (2005). "Iraq: Italian Lessons Learned." Military Review(March-April): 58-61.

Capra, F. (1982). The Turning Point: science, society and the rising culture. New York, Bantam Books.

Capra, F. (2002). The Hidden Connections: A Science for Sustainable Living. New York, Anchor Books.

Castro, A. P. and K. Ettenger (1996). Indigenous Knowledge and Conflict Management: Exploring Local Perspectives and Mechanisms for Dealing with Community Forestry Disputes. Community Forestry Unit. U. N. F. a. A. Organization. Rome, United Nations Food and Agriculture Organization.

Centre, U. N. (2008) "UN-backed container exhibit spotlights plight of sex trafficked victims." UN News Service.

Clark, H., Ed. (2009). People Power: Unarmed Resistance and Global Solidarity. London, Pluto Press.

Coleman, G. (2011) "Anonymous: From the Lulz to Collective Action." The new Everyday: a media commons project.

Coleman, G. (2012). Coding Freedom: The Ethics and Aesthetics of Hacking. Princeton, Princeton University Press.

Coleman, G. (2012) "Everything you know about Anonymous is wrong " AlJazeera.

Coleman, G. (2012) "Our Weirdness Is Free." Triple Canopy.

Cooper, N., M. Turner, et al. (2011). "The end of history and the last liberal peacebuilder: a reply to Roland Paris." Review of International Studies **37**(4): 1995-2007.

Dawisha, A. I. and K. Dawisha (2003). "How to Build a Democratic Iraq." Foreign Affairs **82**(2).

De Souza Santos, B. (2010). Para Descolonizar el Occidente. Mas allá del pensamiento abismal. Buenos Aires, CLACSO.

Easterly, W. (2006). The White Man's Burden: Why the West's Efforts to Aid the Rest have Done So Much Ill and So Little Good. London, Penguin Press.

Fattah, A. A. E. (2011). After Egypt's revolution, I never expected to be back in Mubarak's jails. The Guardian. London.

Fein, P. (2012) "Democracy Is Obsolete." TechPresident.

Firmo-Fontan, V. (2003). The Media and Conflict Prevention: Warning or Monitoring? The Case of Drvar, Bosnia. Before Emergency: Conflict Prevention and the Media. M. Aguirre, F. Ferrandiz and J.-M. Pureza. Bilbao, University of Deusto Press.

Firmo-Fontan, V. (2003). Responses to Human Trafficking: from the Balkans to Afghanistan. The Political Economy of New Slavery. C. Van den Anker. London, Palgrave.

Firmo-Fontan, V. (2004). Abducted, beaten and sold into prostitution: two women's story from an Iraq in turmoil. The Independent. London.

Fisher, R., W. Ury, et al. (1991). Getting to Yes: Negotiating Agreement Without Giving In. London, Penguin Books.

Fisher, S. and L. Zimina (2008). Just Wasting Our Time? Provocative Thoughts for Peacebuilders. Berghof Dialogue Series. Berlin, Berghof Research Centre for Constructive Conflict Management.

Fisk, R. (2002). Pity the Nation: the abduction of Lebanon. New York, Nation Books.

Fontan, V. (2007). Understanding Islamic Terrorism: Humiliation Awareness and the Role for Nonviolence. Nonviolence: An Alternative for Defeating Global Terror(ism). R. Summy and R. Senthil. Hauppauge, Nova Science Publishers.

Fontan, V. (2008). Voices from Post-Saddam Iraq: Living with Terrorism, Insurgency and New Forms of Tyranny. Westport, CT, Praeger Security International.

Fontan, V. (2011) "Gaddafi, Sodomy, and Liberal Peace." Peace and Conflict Monitor.

Foucault, M. (1982). The Archeology of Knowledge & The Discourse on Language. New York, Vintage Books.

Frenkel, S. (2011) "After the Revolution, Arab Women Seek More Rights." National Public Radio.

Galtung, J. (1969). "Violence, Peace and Peace Research." Journal of Peace Research **6**(3): 167-191.

Gardner, H. (2008). Five Minds for the Future. Cambridge, Harvard University Press.

Geneva Declaration (2011). Global Burden of Armed Violence 2011. Geneva, Geneva Declaration on Armed Violence and Development.

Gettelman, J. (2011). Rapes are again reported in Eastern Congo. The New York Times. New York.

Gettleman, J. (2007). Rape Epidemic Raises Trauma of Congo War. The New York Times. New York.

Gilstrap, D. L. (2005). "Strange attractors and human interaction: Leading Complex Organizations through the Use of Metaphors." Complicity: An International Journal of Complexity and Education **2**(1): 55-69.

Gleditsch, N. P. (1992). "Democracy and Peace." Journal of Peace Research **29**(4): 369-376.

Gleick, J. (1998). Chaos: The Amazing Science of the Unpredictable. London, Vintage.

Gonzalez, N. (2009). The Sunni-Shia Conflict: Understanding Sectarian Violence in the Middle East. Santa Ana, CA, Nortia Press.

Gonzalez, P. (2012). "Occupy--A View from the Grassroots: "Thoughts on Diversity, Intersectionality, Strategy, and Movement Building"." Peace Studies Journal **5**(1).

Gorman, S. (2012). Alert on Hacker Power Play. The Wall Street Journal. New York.

Hamzeh, A. N. (2004). In the Path of Hizbullah. Syracuse, Syracuse University Press.

Harvard Humanitarian Initiative (2010). "Now, The World is Without Me": An Investigation of Sexual Violence in Eastern Democratic Republic of Congo. Cambridge, Harvard University.

Hassan, S. (2010) "It's Time to End the Church of Scientology's Tax-Exempt Status." The Huffington Post.

Hendawi, H. (2012) "Mubarak's ex-PM claims to win Egypt president vote " Associated Press.

Hochschild, A. (1999). King Leopold's Ghost: A story of greed, terror, and heroism in colonial Africa. Boston, Mariner Books.

Hock, D. (2005). One from Many: VISA and the Rise of Chaordic Organization. San Francisco, Berrett-Koehler Publishers.

Holbrooke, R. (1998). To End a War. New York, Random House.

Holling, C. S. (2001). "Understanding the Complexity of Economic, Ecological, and Social Systems." Ecosystems **4**: 390-405.

Honeyman, C., J. Coben, et al., Eds. (2009). Rethinking Negotiation Teaching: Innovations and Context and Culture. Saint Paul, CreateSpace.

Honeyman, C., J. Coben, et al., Eds. (2010). Venturing Beyond the Classroom. Saint Paul, CreateSpace.

hooks, b. (1994). Teaching to Transgress. New York, Routledge.

Human Rights Watch (2011). Egypt: Don't Cover Up Military Killing of Copt Protesters. J. Stork. New York, Human Rights Watch.

Amnesty International (2011). Lawless law: Detentions under the Jammu and Kashmir Public Safety Act. London, Amnesty International.

Israel, M. and I. Hay (2006). Research Ethics for Social Scientists. London, Sage.

Jaber, H. (1997). Hezbollah: born with a Vengeance. New York, Columbia University Press.

Johnson, M. (1986). Class and Client in Beirut: the Sunni Muslim community and the Lebanese state, 1840-1985. Reading, Ithaca Press.

Johnson, M. (2001). All Honourable Men: The Social Origins of War in Lebanon. London, I.B. Tauris Publishers.

Johnson, S. (2001). Emergence. New York, Scribner.

Keating, J. E. (2011) "From Tahrir Square to Wall Street." Foreign Policy.

Keen, D. (2008). Complex Emergencies. Cambridge, Polity.

Kennedy, D. (2004). The Dark Sides of Virtue: reassessing international humanitarianism. Princeton, Princeton University Press.

Khomeini, R. (1981). Islam and Revolution: Writings and Declarations of Imam Khomeini. Berkeley, Mizan Press.

Kincheloe, J. and P. McLaren (2007). Critical Pedagogy: Where are we now? New York, Peter Lang.

Korzybski, A. (1931). "A Non-Aristotelian System and its Necessity for Rigour in Mathematics and Physics". paper presented before the meeting of the American Association for the Advancement of Science, December 28, 1931. Reprinted in Science and Sanity, 1933, p. 747–61. New York, Institute for General Semantics.

Kristof, N. D. (2011). 'Three Cups of Tea,' Spilled. The New York Times. New York.

Kristof, N. D. and S. WuDunn (2010). Half the Sky: Turning Oppression into Opportunity for Women Worldwide. New York, Vintage.

Kuhn, T. (1996). The Structure of Scientific Revolutions. Chicago, University of Chicago Press.

Laden, U. b. (2012). Letters from Abbottabad. Harmony Program. C. T. C. a. W. Point. West Point, United States Military Academy.

Lederach, J. P. (1995). Preparing for Peace: Conflict Transformation Across Cultures. Syracuse, Syracuse University Press.

Leigh, D. and L. Harding (2011). WikiLeaks: Inside Julian Assange's War on Secrecy. London, PublicAffairs.

Lindner, E. G. (2006). Making Enemies: Humiliation and International Conflict. Westport, CT, Preager Security International.

Lowenstein, A. K. (2009). The Myth of Normalcy: Impunity and the Judiciary in Kashmir. Human Rights Clinic Yale Law School. New Haven, Yale University.

Lukes, S. (2005). Power: A Radical View. London, Palgrave.

Lynch, C. (2004). U.N. Sexual Abuse Alleged in Congo. The New York Times. New York.

MacGinty, R. (2010). Gilding the lily? International support for indigenous and traditional peacebuilding. Palgrave Advances in Peacebuilding: Critical Developments and Approaches. O. Richmond. Basingstoke, Palgrave Macmillan.

Marder, M. (2012). "Resist like a plant! On the Vegetal Life of Political Movements." Peace Studies Journal **5**(1).

Maren, M. (1997). The Road to Hell: the ravaving effects of foreign aid and international charity. New York, The Free Press.

Meinardus, R. (2011) "Egypt's unfinished revolution." Egypt Daily News.

Miller, P. (2010). Smart Swarm: Using Animal Behaviour to Change our World. London, Collins.

Misra, N. and R. Pandita (2010). The Absent state: Insurgency as an Excuse for Misgovernance. New Delhi, Hachette India.

Morin, E. (1977). La méthode. 1. La Nature de la Nature. Paris, Editions du Seuil.

Moyo, D. (2010). Dead Aid. Why Aid Makes Things Worse and How There is Another Way for Africa. New York, Farrar, Straus and Giroux.

Norton, A. R. (1987). Amal and the Shi'a: Struggle for the South of Lebanon. Austin, University of Texas Press.

Nova Lomax, J. (2010). WikiLeaks: Texas Company Helped Pimp Little Boys to Stoned Afghan Cops. Houston Press. Houston.

Nyhan, B. and J. Reifler (2010). "When Corrections Fail: The Persistance of Political Misperceptions." Political Behaviour **32**(2): 303-330.

O'Dea, J. (2012). Cultivating Peace: Becoming a 21st-Century Peace Ambassador, Shift Books.

Painter, N. I. (2010). The History of White People. New York, Norton.

Palmer Harik, J. (2004). Hezbollah: The Changing Face of Terrorism. New York, I.B. Tauris.

Pateman, C. (1989). The disorder of women: democracy, feminism, and political theory. Palo Alto, Stanford University Press.

Peer, B. (2008). Curfewed Night. New Delhi, Random House India.

Pilkington, E. (2012). Bradley Manning's treatment was cruel and inhuman, UN torture chief rules. The Guardian. London.

Pruitt, D., J. Rubin, et al. (2003). Social Conflict: Escalation, Stalemate and Settlement. New York, McGraw-Hill Humanities.

Redaccion (2012) "Capturan a 13 pandilleros de la MS durante operativo en Mejicanos." La Pagina.

Rhoads, C., G. A. Fowler, et al. (2009) "Iran Cracks Down on Internet Use, Foreign Media." Wall Street Journal.

Richmond, O. (2007). The Transformation of Peace. London, Palgrave, Macmillan.

Richmond, O. (2011). A Post-Liberal Peace. London, Routledge.

Richmond, O. (2012). Failed statebuilding Versus Peace Formation. Manchester, Manchester University.

Rieff, D. (2002). A Bed for the Night: humanitarianism in crisis. New York, Simon and Schuster Paperbacks.

Rumi, J. a.-D., C. Barks, et al. (1995). The Essential Rumi. San Francisco, Harper.

Sankari, J. (2005). Fadlallah: The making of a Radical Shi'ite Leader. London, Saqi Books.

Siebel, J. (2011). Miss Representation. United states of America, ro*co films educational.

Slabodsky, S. (2012). "It is the history, stupid!: A Dialectical Reading of the Utopian Limitations of the US 'Occupy' Movements." Peace Studies Journal **5**(1).

Solis, M. (2010) "Mueren 11 personas calcinadas al interior de bus en Mejicanos." La Prensa Grafica.

Somaiya, R. (2011) "Hackers Shut Down Government Sites." New York Times.

Staff (2008) "Al-Qa'eda in Iraq alienated by cucumber laws and brutality." The Telegraph.

Staff (2011). Somaliland Ready to Aid Famine-Hit Somalia. Somaliland Times. Hargeisa.

Stauth, G. (1991). "Revolution in Spiritless Times. An Essay on Michel Foucault's Enquiries into the Iranian Revolution." International Sociology **6**(3): 259-280.

Stewart, R. (2007). The Prince of the Marshes: And Other Occupational Hazards of a Year in Iraq. Fort Washington, Harvest Books.

Stone, H. (2012) "El Salvador Denies Negotiating with Gangs." In Sight: Organized Crime in the Americas.

Tarek, S. (2011) "Egypt's police after the revolution: Brutality combines with lack of security." Ahram Online.

Targ, R. and J. J. Hurtak (2006). The End of Suffering. Charlottesville, VA, Hampton Roads

Tuhiwai Smith, L. (2008). Decolonizing Methodologies: Research and Indigenous Peoples. New York, Zed Books.

Tulard, J., J.-F. Fayard, et al. (1987). Histoire et dictionnaire de la Révolution française. Paris, Robert Laffont.

United Nations (2000). United Nations Convention against Transnational Organized Crime and its Protocols. U. Nations. Palermo, Italy.

Untersinger, M. (2012) "WikiLeaks et les Anonymous : un mariage de raison." Rue89.

Vaux, T. (2001). The Selfish Altruist: relief work in famine and war. London, Earthscan.

Watch, H. R. (2011). Egypt: Government Moves to Restrict Rights and Democracy Groups. New York, Human Rights Watch.

Wenger, E. (1998). "Communities of Practice: Learning as a Social System." Systems Thinker **9**(5).

Wheatley, M. (2006). Leadership and the new science: Learning about Organization from an orderly perspective. San-Francisco, Berrett-Koehler Publishers.

Wheatley, M. and D. Frieeze (2006). "Using emergence to take social innovationt to scale." Writings.

Notes

1 This paper was later published as: Firmo-Fontan, V. (2003). The Media and Conflict Prevention: Warning or Monitoring? The Case of Drvar, Bosnia. Before Emergency: Conflict Prevention and the Media. M. Aguirre, F. Ferrandiz and J.-M. Pureza. Bilbao, University of Deusto Press.

2 Holbrooke, R. (1998). To End a War. New York, Random House.

3 Firmo-Fontan, V. (2003). The Media and Conflict Prevention: Warning or Monitoring? The Case of Drvar, Bosnia. Before Emergency: Conflict Prevention and the Media. M. Aguirre, F. Ferrandiz and J.-M. Pureza. Bilbao, University of Deusto Press.

4 See http://www.oscebih.org/Default.aspx?id=54&lang=EN; accessed on March 2nd 2011.

5 Interview with Mr. Marceta, Drvar, August 2001.

6 Dayton Peace Agreement, Annex 10, Article II. Annex 10, Article I provided that the civilian aspects of the peace settlements were: the continuation of the humanitarian aid effort so long as required, rehabilitation of the infrastructure and economic reconstruction; establishment of political and institutional institutions in BiH; proportion of respect for human rights and the return of displaced persons and refugees; and the holding of free and fair elections.

7 Unattributable interviews with four OSCE former local and international staff.

8 The following article accounts not only for Ms C.'s appointment, but also for her behavior, endangering Italian troops, during the second battle of Nasiriya in May 2004. Cappelli, R. (2005). "Iraq: Italian Lessons Learned." Military Review (March-April): 58-61.

9 For a detailed account, albeit less controversial, of C.'s work in Iraq, see: Stewart, R. (2007). The Prince of the Marshes: And Other Occupational Hazards of a Year in Iraq. Fort Washington, Harvest Books.

10 http://www.repubblica.it/2008/03/sezioni/cronaca/ndrangheta-1/voti-estero/voti-estero.html; accessed on March 9th, 2011.

11 http://www.itablogs4darfur.blogspot.com/2007/08/nyala-italian-hospital-no-support-from.html; accessed on March 9th 2011.

12 Interview with Jean-Jacques Purusi, Mamas for Africa, Bukavu, April 22nd 2011.

13 Gettelman, J. (2011). Rapes are again reported in Eastern Congo. The New York Times. New York.

14 Un-attributable interviews with two local NGO workers Bukavu; one hotel owner, Uvira; one lawyer, Goma; one nightclub owner, Goma; Several visits to nightclubs in Bukavu, Goma and Uvira witnessing international personnel interactions with minors. See also: Caplan, G. (2012) "Peacekeepers gone wild: How much more abuse will the UN ignore in Congo?" The Globe and Mail., available online: http://www.theglobeandmail.com/news/politics/second-reading/peacekeepers-gone-wild-how-much-more-abuse-will-the-un-ignore-in-congo/article4462151/

15 Interview with Professor Jean-Claude Mubalama Zibona, Catholic University of Bukavu, April 21st 2011, referring to law #06/018 of July 20th 2006, articles 167-170-171bis-174c.

16 Kristof, N. D. (2011). 'Three Cups of Tea,' Spilled. The New York Times. New York.

17 A methodological discussion of both style and research can be found at the end of this chapter.

18 Fontan, V. (2008). Voices from Post-Saddam Iraq: Living with Terrorism, Insurgency and New Forms of Tyranny. Westport, CT, Praeger Security International. This de-Baathification program, inspired from post World War II de-Nazification was the first step

toward liberal/orthodox peace-building in post-Saddam Iraq in 2003.

19 For a fuller account of Huda and Sajeeda's ordeal, see my original article: Firmo-Fontan, V. (2004). Abducted, beaten and sold into prostitution: two women's story from an Iraq in turmoil. The Independent. London.

20 For an interestingly prioritized nation-building/democratization narrative at the time, see Dawisha, A. I. and K. Dawisha (2003). "How to Build a Democratic Iraq." Foreign Affairs **82**(2).

21 While the overhead expenses of Amnesty International USA account for 4.2% of their annual budget, their financial score is one of the lowest in terms of the actual funds that directly benefit people, excluding infrastructure, per diems, luxury accommodation for their staff, high salaries, etc. See: http://www.charitynavigator.org/index.cfm?bay=search.summary&orgid=3294; accessed on February 10th, 2012. For a discussion on the high valedictory payment of Amnesty International's former Secretary general Irene Khan, see: http://www.civilsociety.co.uk/finance/news/content/8390/charity_commission_has_no_jurisdiction_over_board_members_payment_from_amnesty; accessed on Febryary 10th, 2012.

22 It is questionable that Human Trafficking only falls within the remit of organized crime, since it is undeniable that the presence of peace operations and peace-keepers in post-conflict settings increases the demand for sexual services, see: Firmo-Fontan, V. (2003). Responses to Human Trafficking: from the Balkans to Afghanistan. The Political Economy of New Slavery. C. Van den Anker. London, Palgrave. This "parking" of the issue within the remit of Organized Crime absolves peace operations from being probed as to whether they are ensuring any structural basis for human rights abuses in UN missions.

23 United Nations (2000). United Nations Convention against Transnational Organized Crime and its Protocols. United Nations. Palermo, Italy., Annex II, pp. 42-43.

24 See http://www.icrc.org/ihl.nsf/7c4d08d9b287a42141256739003e636b/f6c8b9fee14a77fdc125641e0052b079; accessed on April 24th 2011.

25 Fontan, V. (2008). Voices from Post-Saddam Iraq: Living with Terrorism, Insurgency and New Forms of Tyranny. Westport, CT, Praeger Security International.

26 Korzybski, A. (1931). "A Non-Aristotelian System and its Necessity for Rigour in Mathematics and Physics". American Association for the Advancement of Science, Science and Sanity Commission. New Orleans.

27 Bales, A. and R. Soodalter (2009). The Slave Next Door: human trafficking and slavery in America today. Berkeley, University of California Press.

28 Centre, U. N. (2008) "UN-backed container exhibit spotlights plight of sex trafficked victims." UN News Service.

29 What Abu Baker should have known before arriving is that his hotel belongs to and is managed by Shi'ite Muslims. Amidst the animosity between Sunni and Shi'ite Muslims, exacerbated by the US invasion of the country, it is likely that Abu Baker's treatment at the hands of his management is influenced by his religious identity.

30 Interview with Abdul Karim, liaison with Baghdad immigration office, April 2011.

31 Skype interview with Buk and Koran, April 26th 2011.

32 Interview with Abdul Karim, liaison with Baghdad immigration office, April 2011.

33 Richmond, O. (2007). The Transformation of Peace. London, Palgrave, Macmillan.

34 Nova Lomax, J. (2010). WikiLeaks: Texas Company Helped Pimp Little Boys to Stoned Afghan Cops. Houston Press. Houston.; the cable can be assessed on http://www.guardian.co.uk/world/us-embassy-cables-documents/213720; accessed on March

9th, 2011. Dancing boys in Afghanistan, referred to as Bacha Bazi, are sold or abducted from their families to become the sex slaves of notables, powerful political and religious figures. This human rights abuse was documented in a Public Broadcasting Service documentary: http://www.pbs.org/wgbh/pages/frontline/dancing-boys/; accessed on March 11th 2012.

35 Bolkovac, K. and C. Lynn (2011). The Whistleblower: Sex Trafficking, Military Contractors and One Woman's Fight for Justice. London, Palgrave Macmillan.; Firmo-Fontan, V. (2003). Responses to Human Trafficking: from the Balkans to Afghanistan. The Political Economy of New Slavery. C. Van den Anker. London, Palgrave.

36 Bolkovac, K. and C. Lynn (2011). The Whistleblower: Sex Trafficking, Military Contractors and One Woman's Fight for Justice. London, Palgrave Macmillan.

37 See http://www.dyn-intl.com/; accessed on March 10th, 2011.

38 Anderson, M. B. (1999). Do No Harm: How Aid can Support Peace – Or War. Boulder, Lynne Rienner Publishers.

39 Moyo, D. (2010). Dead Aid. Why Aid Makes Things Worse and How There is Another Way for Africa. New York, Farrar, Straus and Giroux.

40 Easterly, W. (2006). The White Man's Burden: Why the West's Efforts to Aid the Rest have Done So Much Ill and So Little Good. London, Penguin Press.

41 White here refers to a paradigm of social whiteness. For a discussion on the whiteness paradigm and the "worship" of whiteness for economic, social, and political aims, see Painter, N. I. (2010). The History of White People. New York, Norton.

42 Brandstetter, R. H. and V. C. Fontan (2005). Final Report for Political Process Assistance Review. Monitoring and Evaluation Performance Program, Phase II (MEPP II). I. International Business & Technical Consultants. Washington, D.C., United States Agency for International Development.

43 Ibid.

44 Keen, D. (2008). Complex Emergencies. Cambridge, Polity., specifically Chapter 6.

45 For a legal and international law debate on humanitarian intervention, see Kennedy, D. (2004). The Dark Sides of Virtue: reassessing international humanitarianism. Princeton, Princeton University Press.

46 Caplan, G. (2009). Obama and Africa – a major disappointment. The Nation. New York.

47 Easterly, W. (2006). The White Man's Burden: Why the West's Efforts to Aid the Rest have Done So Much Ill and So Little Good. London, Penguin Press.; see also Hochschild, A. (1999). King Leopold's Ghost: A story of greed, terror, and heroism in colonial Africa. Boston, Mariner Books.

48 Richmond, O. (2007). The Transformation of Peace. London, Palgrave, Macmillan.

49 Rumi, J. a.-D., C. Barks, et al. (1995). The Essential Rumi. San Francisco, Harper, p. 36.

50 Maren, M. (1997). The Road to Hell: the ravaging effects of foreign aid and international charity. New York, The Free Press.; Rieff, D. (2002). A Bed for the Night: humanitarianism in crisis. New York, Simon and Schuster Paperbacks.

51 Fisher, R., W. Ury, et al. (1991). Getting to Yes: Negotiating Agreement Without Giving In. London, Penguin Books.; Pruitt, D., J. Rubin, et al. (2003). Social Conflict: Escalation, Stalemate and Settlement. New York, McGraw-Hill Humanities.

52 For a dialogue on the local/indigenous and universal, see: Castro, A. P. and K. Ettenger (1996). Indigenous Knowledge and Conflict Management: Exploring Local Perspectives and Mechanisms for Dealing with Community Forestry Disputes. Community Forestry Unit. U. N. F. a. A. Organization. Rome, United Nations Food and Agriculture Organization. See also MacGinty, R. (2010). Gilding the lily? International support for indigenous and

traditional peacebuilding. Palgrave Advances in Peacebuilding: Critical Developments and Approaches. O. Richmond. Basingstoke, Palgrave Macmillan.
It is it is worth noting that ever since Getting to Yes, Rubin, Pruitt & Kim, there have been many critical voices both on the theory and pedagogy of the conflict resolution field, even from within the "Northern" scholars and practitioners' crowds. However, the basis of teaching negotiation often remains the same, while multicultural and gender aspects are provided as an addition and not a basis. For some very relevant critiques, see: Honeyman, C., J. Coben, et al., Eds. (2009). Rethinking Negotiation Teaching: Innovations and Context and Culture. Saint Paul, CreateSpace., and Honeyman, C., J. Coben, et al., Eds. (2010). Venturing Beyond the Classroom. Saint Paul, CreateSpace.

53 Easterly, W. (2006). The White Man's Burden: Why the West's Efforts to Aid the Rest have Done So Much Ill and So Little Good. London, Penguin Press.

54 Richmond, O. (2012). Failed Statebuilding Versus Peace Formation. Manchester, Manchester University.

55 Galtung, J. (1969). "Violence, Peace and Peace Research." Journal of Peace Research **6**(3): 167-191.

56 For a vivid analysis of this particular narrative applied to Rwanda, Somalia and Bosnia-Herzegovina, see: Rieff, D. (2002). A Bed for the Night: humanitarianism in crisis. New York, Simon and Schuster Paperbacks., Maren, M. (1997). The Road to Hell: the ravaving effects of foreign aid and international charity. New York, The Free Press., and Caplan, G. (2008). The Betrayal of Africa. Toronto, Groundwood.

57 For an excellent debate on the issue, see Caplan, G. (2008). The Betrayal of Africa. Toronto, Groundwood.

58 For an illustration of how the liberal peace paradigm directly costs civilian lives, see chapter five of my Iraq book: Fontan, V. (2008). Voices from Post-Saddam Iraq: Living with Terrorism,

Insurgency and New Forms of Tyranny. Westport, CT, Praeger Security International.

59 Ibid.

60 Capra, F. (1982). The Turning Point: science, society and the rising culture. New York, Bantam Books.

61 Ibid.

62 Kuhn, T. (1996). The Structure of Scientific Revolutions. Chicago, University of Chicago Press.

63 Bolkovac, K. and C. Lynn (2011). The Whistleblower: Sex Trafficking, Military Contractors and One Woman's Fight for Justice. London, Palgrave Macmillan.

64 Bauman, Z. (2001). Modernity and the Holocaust. Ithaca, Cornell University Press.

65 Achebe's novel *Things Fall Apart*, for instance, illustrates the deeply rooted patriarchal issues of pre-colonial Nigeria. This argument in no way means to return to an idealized antique order of social inequalities. Achebe, C. (1994). Things Fall Apart. London, Anchor Books.

66 While bringing indigenous issues at the forefront of the social debates both in Guatemala and worldwide, Menchu's presidential bid in 2007 and 2011 only gathered 3% of votes, signifying a shift in popular support among indigenous communities that saw her being co-opted by an institutional/elitist power-base.

67 Gettleman, J. (2007). Rape Epidemic Raises Trauma of Congo War. The New York Times. New York.; Initiative, H. H. (2010). "Now, The World is Without Me": An Investigation of Sexual Violence in Eastern Democratic Republic of Congo. Cambridge, Harvard University.

68 Israel, M. and I. Hay (2006). Research Ethics for Social Scientists. London, Sage.

69 Kincheloe, J. and P. McLaren (2007). Critical Pedagogy: Where are we now? New York, Peter Lang.; Hooks, B. (1994). Teaching to Transgress. New York, Routledge.

70 Tuhiwai Smith, L. (2008). Decolonizing Methodologies: Research and Indigenous Peoples. New York, Zed Books.

71 On former MONUC abuses, see: Lynch, C. (2004). U.N. Sexual Abuse Alleged in Congo. The New York Times. New York.

72 Interview with Dr. Jean-Jacques Purusi Sadiki, Bukavu, April 24th 2011.

73 The interviews of Geovanni Morales and Father Antoño Rodriguez Lopez that are the basis for this analysis took place between the 11th and the 20th of March 2012. Most of the information was collected by a group of students and affiliates of the MA in Media, peace and conflict studies at the University for Peace: Hiroko Awano, Atkilt Geleta, Lucian Segura, and Sandra Sharman. I met with Father Antoño the day of Geovanni's arrest, and was in contact with Geovanni only through e-mail communication.

74 Solis, M. (2010) "Mueren 11 personas calcinadas al interior de bus en Mejicanos." La Prensa Grafica; accessed online on April 23rd, 2012: http://www.laprensagrafica.com/el-salvador/judicial/126811-mueren-11-personas-calcinadas-al-interior-de-bus-en-mejicanos.html

75 Redaccion (2012) "Capturan a 13 pandilleros de la MS durante operativo en Mejicanos." La Pagina; accessed online on April 23rd, 2012: http://www.lapagina.com.sv/nacionales/64119/2012/03/20/Capturan-a-13-pandilleros-de-la-MS-durante-operativo-en-Mejicanos

76 Stone, H. (2012) "El Salvador Denies Negotiating with Gangs." In Sight: Organized Crime in the Americas; accessed online on April 23rd, 2012: http://insightcrime.org/insight-latest-news/item/2371-el-salvador-denies-negotiating-with-gangs

77 See http://www.lapagina.com.sv/nacionales/64157/2012/03/20/Carta-del-padre-Rodriguez-ante-supuestas-negociaciones-del-gobierno-con-pandillas; accessed online on April 23rd 2012.

78 Telephone interview between Lucian Segura and Ingrid Saravia, Father Toño's administrative assistant, March 2012.

79 Clark, H., Ed. (2009). People Power: Unarmed Resistance and Global Solidarity. London, Pluto Press.

80 Fattah, A. A. E. (2011). After Egypt's revolution, I never expected to be back in Mubarak's jails. The Guardian. London; accessed online on November 4th 2011: http://www.guardian.co.uk/commentisfree/2011/nov/02/egypt-revolution-back-mubarak-jails

81 Frenkel, S. (2011) "After the Revolution, Arab Women Seek More Rights." National Public Radio; accessed online on November 4th 2011: http://www.npr.org/2011/08/06/137482442/after-the-revolution-arab-women-seek-more-rights

82 Watch, H. R. (2011). Egypt: Don't Cover Up Military Killing of Copt Protesters. J. Stork. New York, Human Rights Watch; accessed online on November 4th 2011: http://www.hrw.org/news/2011/10/25/egypt-don-t-cover-military-killing-copt-protesters

83 Tarek, S. (2011) "Egypt's police after the revolution: Brutality combines with lack of security." Ahram Online; accessed online on November 4th 2011: http://english.ahram.org.eg/NewsContent/1/64/25124/Egypt/Politics-/Egypt%E2%80%99s-police-after-the-revolution-brutality-comb.aspx

84 Watch, H. R. (2011). Egypt: Government Moves to Restrict Rights and Democracy Groups. New York, Human Rights Watch; accessed online on November 4th 2011: http://www.hrw.org/news/2011/09/26/egypt-government-moves-restrict-rights-and-democracy-groups

85 Meinardus, R. (2011) "Egypt's unfinished revolution." Egypt Daily News; accessed online on November 4th 2011: http://www.

thedailynewsegypt.com/columnists/egypts-unfinished-revolution.html

86 Kuhn, T. (1996). The Structure of Scientific Revolutions. Chicago, University of Chicago Press.

87 Nyhan, B. and J. Reifler (2010). "When Corrections Fail: The Persistance of Political Misperceptions." Political Behaviour **32**(2): 303-330.

88 Pateman, C. (1989). The disorder of women: democracy, feminism, and political theory. Palo Alto, Stanford University Press.

89 Rumi, J. a.-D., C. Barks, et al. (1995). The Essential Rumi. San Francisco, Harper.

90 Boff, L. and M. Hathaway (2009). The Tao of Liberation, Exploring the Ecology of Transformation. Maryknoll, Orbis Books.

91 Siebel, J. (2011). Miss Representation. United states of America, ro*co films educational.

92 On the Aristotelian excluded middle, see Targ, R. and J. J. Hurtak (2006). The End of Suffering. Charlottesville, VA, Hampton Roads Publishing Company, Inc.

93 Ibid.

94 Quoted in: Bateson, G. (2000). Form, Substance, and Difference. Steps to an Ecology of the Mind. G. Bateson. Chicago, University of Chicago Press, p. 455.

95 Declaration, G. (2011). Global Burden of Armed Violence 2011. Geneva, Geneva Declaration on Armed Violence and Development.

96 Ibid.

97 Interview with Professor Victor Valle, University for Peace, Costa Rica, November 4th 2011.

98 Capra, F. (1982). The Turning Point: science, society and the rising culture. New York, Bantam Books.

99 Easterly, W. (2006). The White Man's Burden: Why the West's Efforts to Aid the Rest have Done So Much Ill and So Little Good. London, Penguin Press.; Richmond, O. (2007). The Transformation of Peace. London, Palgrave, Macmillan.

100 Interview with Bano Hameeda, Srinagar, India, November 2008.

101 Bose, S. (2003). Kashmir: Roots of Conflict, Paths to Peace. Cambridge, Harvard University Press.

102 Ibid.

103 These figures are given respectively by the Indian state, 50,000 casualties, and diverse Kashmiri Human Rights groups, 70,000. They were gathered in an interview with Wajahat Ahmad, Srinagar, India, November 2010.

104 In India, the federal version of the well-known US Patriot Act is the Prevention of Terrorism Act. Each state has then its own application of it, which in Kashmir has been pushed to an extreme in the creation of the Army Special Forces Act, which gives the army the right to any intervention deemed necessary to maintain "civil peace" in Kashmir. This in effect give the Indian Army a *carte blanche* to arrest, interrogate, and detain anyone suspected of carrying out actions against the sovereignty of the Indian state. See: Amnesty International (2011). Lawless law: Detentions under the Jammu and Kashmir Public Safety Act. London, Amnesty International.

105 Misra, N. and R. Pandita (2010). The Absent State: Insurgency as an Excuse for Misgovernance. New Delhi, Hachette India.

106 Lowenstein, A. K. (2009). The Myth of Normalcy: Impunity and the Judiciary in Kashmir. Human Rights Clinic Yale Law School. New Haven, Yale University.

107 Gleditsch, N. P. (1992). "Democracy and Peace." Journal of Peace Research **29**(4): 369-376.

108 Fisher, S. and L. Zimina (2008). Just Wasting Our Time? Provocative Thoughts for Peacebuilders. Berghof Dialogue

Series. Berlin, Berghof Research Centre for Constructive Conflict Management.

109 Ibid.; Richmond, O. (2007). The Transformation of Peace. London, Palgrave, Macmillan.; Lederach, J. P. (1995). Preparing for Peace: Conflict Transformation Across Cultures. Syracuse, Syracuse University Press.

110 The details of many NGOs or media outlets of the Hezbollah are freely accessible on the Internet, see Jihad al Binaa http://www.yellowpages.com.lb/qdo_search.php?q=organization&x=0&y=0&any=all&p=38; and Al Manar Television http://www.yellowpages.com.lb/qdo_search.php?q=al+manar&x=53&y=25&any=all; accessed on May 21st 2012.

111 For more information on the Hezbollah's relationship with international organizations, and specifically its policy shift on the topic between 1985 and now, see: Alagha, J. E. (2006). The Shifts in Hizbullah's Ideology: Religious Ideology, Political Ideology, and Political Program. Amsterdam, Amsterdam University Press, pp. 196 and 364-365

112 Sankari, J. (2005). Fadlallah: The making of a Radical Shi'ite Leader. London, Saqi Books, p. 198.

113 Hamzeh, A. N. (2004). In the Path of Hizbullah. Syracuse, Syracuse University Press.; interview with Hussein Naboulsi, Hezbollah press Office, summer of 2001.

114 The division between Shi'i and Sunni Muslims came after the death of Prophet Mohammed in 632, when part of the Muslim community, the Shi'i, chose his cousin and son-in-law, Ali, to be his successor, while others, the Sunnis, looked to his political companion Abu Bakr as his rightful successor.

115 Sankari, J. (2005). Fadlallah: The making of a Radical Shi'ite Leader. London, Saqi Books.

116 Johnson, M. (2001). All Honourable Men: The Social Origins of War in Lebanon. London, I.B. Tauris Publishers.

117 Hamzeh, A. N. (2004). In the Path of Hizbullah. Syracuse, Syracuse University Press.

118 It is important to note that the name Hezbollah had already been named in several political declarations as early as in 1984. Yet the name Hezbollah only became prominent, thus visible, in Western eyes, with the dissemination of its 1985 Manifesto. See: Alagha, J. E. (2006). The Shifts in Hizbullah's Ideology: Religious Ideology, Political Ideology, and Political Program. Amsterdam, Amsterdam University Press, p. 35.

119 Fisk, R. (2002). Pity the Nation: the abduction of Lebanon. New York, Nation Books.

120 Sankari, J. (2005). Fadlallah: The making of a Radical Shi'ite Leader. London, Saqi Books.

121 Fisk, R. (2002). Pity the Nation: the abduction of Lebanon. New York, Nation Books.

122 Bowyer Bell, J. (1996). Terror out of Zion: The Fight for Israeli Independence. New Brunswick, Transaction Publishers.

123 For mainstream media documentaries, routinely called "Inside the Hezbollah", see: http://www.youtube.com/watch?feature=player_embedded&v=NlU_edvJZk4; accessed on May 23rd 2012, http://www.youtube.com/watch?feature=player_embedded&v=Ibm9B9Ht-m0; accessed on May 23rd 2012, http://www.fipa.tm.fr/fr/programmes/2003/al-manar-tv-au-nom-du-hezbollah-6588.htm; accessed on May 23rd 2012; for academic publications, see: Azani, E. (2008). Hezbollah: The Story of the Party of God. London, Palgrave.; Jaber, H. (1997). Hezbollah: born with a Vengeance. New York, Columbia University Press.; Blanford, N. (2011). Warriors of God: Inside Hezbollah's Thirty-Year Struggle against Israel. New York, Random House.

124 Foucault, M. (1982). The Archeology of Knowledge & The Discourse on Language. New York, Vintage Books.

125 Hamzeh, A. N. (2004). In the Path of Hizbullah. Syracuse, Syracuse University Press.

126 Alagha, J. E. (2011). Hisbullah's Identity Construction. Amsterdam, Amsterdam University Press, p. 185.

127 Gilstrap, D. L. (2005). "Strange attractors and human interaction: Leading Complex Organizations through the Use of Metaphors." Complicity: An International Journal of Complexity and Education **2**(1): 55-69.

128 Ibid, p. 59.

129 Briggs, J. and F. D. Peat (2000). Seven Life Lessons of Chaos: Spiritual Wisdom from the Science of Change. New York, Harper Perenial, p. 15.

130 Keen, D. (2008). Complex Emergencies. Cambridge, Polity.

131 Gilstrap, D. L. (2005). "Strange attractors and human interaction: Leading Complex Organizations through the Use of Metaphors." Complicity: An International Journal of Complexity and Education **2**(1): 55-69, p. 60.

132 Briggs, J. and F. D. Peat (2000). Seven Life Lessons of Chaos: Spiritual Wisdom from the Science of Change. New York, Harper Perenial.

133 Gleick, J. (1998). Chaos: The Amazing Science of the Unpredictable. London, Vintage.

134 Hock, D. (2005). One from Many: VISA and the Rise of Chaordic Organization. San Francisco, Berrett-Koehler Publishers.

135 Wheatley, M. (2006). Leadership and the new science: Learning about Organization from an orderly perspective. San-Francisco, Berrett-Koehler Publishers.

136 Hock, D. (2005). One from Many: VISA and the Rise of Chaordic Organization. San Francisco, Berrett-Koehler Publishers.; and Gleick, J. (1998). Chaos: The Amazing Science of the Unpredictable. London, Vintage.

137 Allah, H. (1985). Nass al-risala al-maftuha allati wajjaha hizb allah ila al-mustad 'afin fi lubnan wa al-'alam. Amal and the Shi'a:

Struggle for the South of Lebanon. A. R. Norton. Austin, University of Texas press.

138 Ibid, p. 173.

139 Johnson, M. (1986). Class and Client in Beirut: the Sunni Muslim community and the Lebanese state, 1840-1985. Reading, Ithaca Press.

140 Norton, A. R. (1987). Amal and the Shi'a: Struggle for the South of Lebanon. Austin, University of Texas Press.

141 Khomeini, I. (1981). Islam and Revolution: Writings and Declarations of Imam Khomeini. Berkeley, Mizan Press.

142 Alagha, J. E. (2010). "Wilayat al-Faqih and Hizbullah's Relations with Iran." Journal of Arabic and Islamic Studies **10**: 24-44.

143 Sankari, J. (2005). Fadlallah: The making of a Radical Shi'ite Leader. London, Saqi Books, p. 176.

144 Alagha, J. E. (2010). "Wilayat al-Faqih and Hizbullah's Relations with Iran." Journal of Arabic and Islamic Studies **10**: 24-44.

145 Jaber, H. (1997). Hezbollah: born with a Vengeance. New York, Columbia University Press, p. 53.

146 Palmer Harik, J. (2004). Hezbollah: The Changing Face of Terrorism. New York, I.B. Tauris.

147 Alagha, J. E. (2010). "Wilayat al-Faqih and Hizbullah's Relations with Iran." Journal of Arabic and Islamic Studies **10**: 24-44, p. 26.

148 *Khum* represents one fifth of one's annual income. As a source on Hezbollah's finances, read: Hamzeh, A. N. (2004). In the Path of Hizbullah. Syracuse, Syracuse University Press.

149 Information gathered in participant observation periods between the summer of 2001 and early 2003.

150 Alagha, J. E. (2011). Hizbullah's Documents: From the 1985 Open Letter to the 2009 Manifesto. Amsterdam, Amsterdam University Press, p. 115.

151 For political reasons, there has not been a population census carried out in Lebanon since 1932, however, it is believed that the Shi'i population in Lebanon has now largely outgrown all other sectarian communities.

152 Alagha, J. E. (2006). *The Shifts in Hizbullah's Ideology: Religious Ideology, Political Ideology, and Political Program*. Amsterdam, Amsterdam University Press.

153 Interviews with various Party officials between 2001 and 2012.

154 Gonzalez, N. (2009). *The Sunni-Shia Conflict: Understanding Sectarian Violence in the Middle East*. Santa Ana, CA, Nortia Press.

155 Fontan, V. (2007). Understanding Islamic Terrorism: Humiliation Awareness and the Role for Nonviolence. *Nonviolence: An Alternative for Defeating Global Terror(ism)*. R. Summy and R. Senthil. Hauppauge, Nova Science Publishers.

156 Laden, U. b. (2012). Letters from Abbottabad. *Harmony Program*. C. T. C. a. W. Point. West Point, United States Military Academy.

157 Alagha, J. E. (2010). "Wilayat al-Faqih and Hizbullah's Relations with Iran." *Journal of Arabic and Islamic Studies* **10**: 24-44.

158 Stauth, G. (1991). "Revolution in Spiritless Times. An Essay on Michel Foucault's Enquiries into the Iranian Revolution." *International Sociology* **6**(3): 259-280.

159 Alagha, J. E. (2010). "Wilayat al-Faqih and Hizbullah's Relations with Iran." *Journal of Arabic and Islamic Studies* **10**: 24-44, p. 27.

160 Staff (2008) "Al-Qa'eda in Iraq alienated by cucumber laws and brutality." *The Telegraph*.

161 Fontan, V. (2008). *Voices from Post-Saddam Iraq: Living with Terrorism, Insurgency and New Forms of Tyranny*. Westport, CT, Praeger Security International.

162 Laden, U. b. (2012). Letters from Abbottabad. Harmony Program. C. T. C. a. W. Point. West Point, United States Military Academy.

163 A 36-hour long interview took place with Sampat Pal Devi and Jay Prakash on April 10th and 11th 2009, in the towns of Badaosa and Attara, Uttar Pradesh, India. While Sampat herself related all the events written about in this chapter to me, I first read a fuller version of them in a biography that was written in French in 2008, and sold widely in several countries. Sampat claims that she only received 6000 euros from the authors for selling her life story, and now realizes that she ought to have received a lot more. She is not bitter about it. This book is, to date, the most detailed recollection of her life: Pal, S. and A. Berthod (2008). Moi, Sampat Pal : chef de gang en sari rose. Paris, Oh! Editions.

164 Coined by Holling and Gunderson, panarchy comes from the "melded image of the Greek god Pan as the epitoma of unpredictable change with the notion of hierarchies across scales." See Holling, C. S. (2001). "Understanding the Complexity of Economic, Ecological, and Social Systems." Ecosystems **4**: 390-405, p. 396.

165 Ibid, p. 392.

166 Ibid, p. 394.

167 Ibid.

168 Ibid, p. 397.

169 Richmond, O. (2011). A Post-Liberal Peace. London, Routledge.

170 Tulard, J., J.-F. Fayard, et al. (1987). Histoire et dictionnaire de la Révolution française. Paris, Robert Laffont.

171 Richmond, O. (2012). Failed Statebuilding Versus Peace Formation. Manchester, Manchester University, p. 14.

172 From this perspective, the Hezbollah as a complex adaptive system is also sustainable within a larger context, hence the fact

that its participation in the Lebanese democratic system allowed for it to reach a vortex formation.

173 Easterly, W. (2006). The White Man's Burden: Why the West's Efforts to Aid the Rest have Done So Much Ill and So Little Good. London, Penguin Press.

174 For an interesting account of the damaging effects of foreign aid and the complexities of North-South relationships to violence, war and peace, see (respectively): Maren, M. (1997). The Road to Hell: the ravaging effects of foreign aid and international charity. New York, The Free Press., and Keen, D. (2008). Complex Emergencies. Cambridge, Polity.

175 Cooper, N., M. Turner, et al. (2011). "The end of history and the last liberal peaebuilder: a reply to Roland Paris." Review of International Studies **37**(4): 1995-2007.

176 De Souza Santos, B. (2010). Para Descolonizar el Occidente. Mas allá del pensamiento abismal. Buenos Aires, CLACSO.

177 See: http://topics.nytimes.com/topics/reference/timestopics/subjects/f/famine/index.html; accessed on June 8th 2012.

178 Staff (2011). Somaliland Ready to Aid Famine-Hit Somalia. Somaliland Times. Hargeisa., available online: http://www.somalilandtimes.net/sl/2011/498/10.shtml; accessed on June 8th, 2012.

179 Bateson, G. (1972). The Cybernetics of "Self": A Theory of Alcoholism. Steps to an Ecology of the Mind. G. Bateson. Chicago, University of Chicago Press, p. 310.

180 Ibid, p. 323.

181 Ibid, p. 323.

182 Ibid, p. 318.

183 Ibid, p. 315.

184 Ibid, p. 324.

185 Bateson, G. (1972). From Versailles to Cybernetics. Steps to an Ecology of the Mind. G. Bateson. Chicago, The University of Chicago Press.

186 Global scholar Evelin Lindner makes a similar assertion in terms of how the humiliation of the German people in the Versailles Treaty led to the democratic election of Adolf Hitler, and subsequently World War II. See: Lindner, E. G. (2006). Making Enemies: Humiliation and International Conflict Westport, CT, Preager Security International, ibid.

187 Bateson, G. (1972). From Versailles to Cybernetics. Steps to an Ecology of the Mind. G. Bateson. Chicago, The University of Chicago Press, p. 481.

188 Fontan, V. (2008). Voices from Post-Saddam Iraq: Living with Terrorism, Insurgency and New Forms of Tyranny. Westport, CT, Praeger Security International, pp. 72-73.

189 See Keating, J. E. (2011) "From Tahrir Square to Wall Street." Foreign Policy., and Barnard, A. (2011) "Occupy Wall Street Meets Tahrir Square." New York Times.

190 Marder, M. (2012). "Resist like a plant! On the Vegetal Life of Political Movements." Peace Studies Journal **5**(1), p. 3.

191 Gonzalez, P. Ibid."Occupy – A View from the Grassroots: "Thoughts on Diversity, Intersectionality, Strategy, and Movement Building""

192 Ibid.

193 Slabodsky, S. Ibid."It is the history, stupid!: A Dialectical Reading of the Utopian Limitations of the US 'Occupy' Movements."

194 O'Dea, J. (2012). Cultivating Peace: Becoming a 21st-Century Peace Ambassador, Shift Books.

195 Bolte Taylor, J. (2006). My Stroke of Insight. New York, Plume, p. 27-28.

196 Ibid, p. 31.

197 Richmond, O. (2007). The Transformation of Peace. London, Palgrave, Macmillan., see Chapter 6.

198 Bolte Taylor, J. (2006). My Stroke of Insight. New York, Plume, p. 29.

199 Ibid, p. 30.

200 Ibid, p. 30 and p. 28.

201 Bateson, G. (1991). Ecology of the Mind: The Sacred. A Sacred Unity: Further Steps to an Ecology of the Mind. R. E. Donaldson. San Francisco, Cornelia & Michael Bessie Book, p. 267.

202 Vaux, T. (2001). The Selfish Altruist: relief work in famine and war. London, Earthscan.

203 Briganti (de), G. (2012) "The Real Reasons for Rafale's Indian Victory." defense-aerospace.com., available online: http://www.defense-aerospace.com/article-view/feature/132379/why-rafale-won-in-india.html; accessed on June 12th, 2012.

204 Fontan, V. (2011) "Gaddafi, Sodomy, and Liberal Peace." Peace and Conflict Monitor.

205 Gardner, H. (2008). Five Minds for the Future. Cambridge, Harvard University Press.

206 Boff, L. (2008). Essential Care: An Ethics of Human Nature. Waco, TX, Baylor University Press, p. 14.

207 Richmond, O. (2007). The Transformation of Peace. London, Palgrave, Macmillan, p. 202.

208 Interviews with Hiro Adam Diriye Sama Wada, November 2009, Hargeisa, Somaliland.

209 Interviews with Edna Adan Ismael, November 2009, International Airport, Djibouti; and Hargeisa, Somaliland.

210 As explained to me by Edna, Somaliland is a former English "protectorate" that obtained its independence in early June 1960. It was joined by Italian Somalia on June 26th 1960. Both joined "protectorates" then chose to be called "Somalia" thereafter, a stra-

tegic mistake on part of Somaliland, the first-born independent republic, as its present claim to political autonomy is understood by the international community as a call for secession and partition. For more information, see: Bulhan, H. A. (2008). The Politics of Cain: One Hundred Years of Crises in Somali Politics and Society. Bethesda, Tayosan International Publishing.

211 Briggs, J. and F. D. Peat (2000). Seven Life Lessons of Chaos: Spiritual Wisdom from the Science of Change. New York, Harper Perenial.

212 Kristof, N. D. and S. WuDunn (2010). Half the Sky: Turning Oppression into Opportunity for Women Worldwide. New York, Vintage.; see also: http://www.youtube.com/watch?feature=player_embedded&v=hfvFoaCrgUo; accessed on June 13th 2012.

213 Somaiya, R. (2011) "Hackers Shut Down Government Sites." New York Times.; see also: Operation Egypt's press release: http://www.youtube.com/watch?v=yOLc3B2V4AM; accessed on June 15th, 2012.

214 IRC private chat with Anon1. Throughout this Chapter, I will not disclose the "nicks", nicknames, of my interlocutors, for privacy as well as security reasons. I will refer to all of my interlocutors as Anon, and will only differentiate them with cardinal numbers. Since Telecomix operates in an open fashion, however, I will refer to him by his real name.

215 Referring to "the map is not the territory", taking live examples and bringing it to our theoretical understanding

216 For an interesting discussion on ant colonies, see: Johnson, S. (2001). Emergence. New York, Scribner.

217 Wenger, E. (1998). "Communities of Practice: Learning as a Social System." Systems Thinker **9**(5).

218 Ibid, p. 2., available online: http://www.ewenger.com/pub/index.htm; accessed on June 15th, 2012.

219 Ibid, p. 2.

220 Capra, F. (2002). The Hidden Connections: A Science for Sustainable Living. New York, Anchor Books, p. 112.

221 Ibid.

222 For a detailed explanation of the Yin-Yang Cosmology, see Morin, E. (1977). La méthode. 1. La Nature de la Nature. Paris, Editions du Seuil, p. 228.

223 Live interview with "nOpants", Costa Rica, December 2011. See also: http://blog.wearpants.org/hacking-for-freedom/; accessed on June 14th 2012.

224 See: http://www.youtube.com/watch?v=UFBZ_uAbxS0; accessed on June 15th, 2012.

225 From the perspective of this book, of course, it can be seen as the tip of the iceberg of the Western "help" addiction, see Chapter IV.

226 Coleman, G. (2011) "Anonymous: From the Lulz to Collective Action." The new Everyday: a media commons project.

227 See http://www.youtube.com/watch?v=JCbKv9yiLiQ; accessed on June 15th, 2012.

228 See https://whyweprotest.net/anonymous-scientology/; accessed on June 15th, 2012.

229 Hassan, S. (2010) "It's Time to End the Church of Scientology's Tax-Exempt Status." The Huffington Post 2012.

230 Untersinger, M. (2012) "WikiLeaks et les Anonymous : un mariage de raison." Rue89.; available online and accessed on June 16th, 2012: http://www.rue89.com/2012/03/04/wikileaks-et-les-anonymous-un-mariage-de-raison-229750

231 Private discussions on #OpTeach with Anon1 and Anon2 on #OpIran, see also Coleman, G. (2012) "Everything you know about Anonymous is wrong " AlJazeera.

232 4Chan, established in the early 2000s was originally an anonymous message board for Japanese manga lovers. It quickly became

a more open forum of Internet subculture. See also Coleman, G. (2012) "Our Weirdness Is Free." Triple Canopy.

233 Rhoads, C., G. A. Fowler, et al. (2009) "Iran Cracks Down on Internet Use, Foreign Media " Wall Street Journal.; available online http://online.wsj.com/article/SB124519888117821213.html; accessed on June 16th 2009.

234 See the contribution of HiveLoic: http://thehackernews.com/2011/02/operation-iran-opiran-anonymous.html, linking to a detail of forthcoming actions: http://piratepad.net/DpQd98grbU, and a supporting video: http://www.youtube.com/watch?v=3_IHeYjiFlY (graphic content); all websites were accessed on June 16th, 2012.

235 For an illustration of this in nature, see: Miller, P. (2010). Smart Swarm: Using Animal Behaviour to Change our World. London, Collins.

236 Ibid.

237 E-mail communication with Pete Fein.

238 Leigh, D. and L. Harding (2011). WikiLeaks: Inside Julian Assange's War on Secrecy. London, PublicAffairs.

239 Pilkington, E. (2012). Bradley Manning's treatment was cruel and inhuman, UN torture chief rules. The Guardian. London.; available online: http://www.guardian.co.uk/world/2012/mar/12/bradley-manning-cruel-inhuman-treatment-un; accessed on June 17th, 2012.

240 Gorman, S. (2012). Alert on Hacker Power Play The Wall Street Journal. New York.; available online: http://online.wsj.com/article_email/SB10001424052970204059804577229390105521090-lMyQjAxMTAyMDIwMDEyNDAyWj.html; accessed on June 18th, 2012.

241 Fein, P. (2012) "Democracy Is Obsolete." TechPresident.; available online: http://techpresident.com/news/22332/op-ed-peter-fein-pdf12-democracy-obsolete; accessed on June 18th, 2012.

242 Wheatley, M. and D. Frieeze (2006). "Using emergence to take social innovation to scale." Writings.; available online: http://www.margaretwheatley.com/articles/emergence.html; accessed on June 18th, 2012.

243 Laden, U. b. (2012). Letters from Abbottabad. Harmony Program. C. T. C. a. W. Point. West Point, United States Military Academy.; SOCOM-2012-0000010-HT, page #1.

244 Hendawi, H. (2012) "Mubarak's ex-PM claims to win Egypt president vote." Associated Press., available online: http://hosted.ap.org/dynamic/stories/M/ML_EGYPT?SITE=AP&SECTION=HOME&TEMPLATE=DEFAULT&CTIME=2012-06-17-23-11-03; accessed on June 19th, 2012.

245 Un-attributable interview of al-Qaeda in Iraq members present in meetings with the Turkish intelligence in the North of Iraq, as well as meetings with the Egyptian Muslim Brotherhood both in Jordan and in Sudan, all within a time span ranging from September 2011 to the time of writing, June 2012.

246 Fein, P. (2012) "Democracy Is Obsolete." TechPresident.

247 Bohm, D. (1996). On Dialogue. London, Routledge.

248 Lukes, S. (2005). Power: A Radical View. London, Palgrave.

249 Bohm, D. (1996). On Dialogue. London, Routledge.

250 Coleman, G. (2012). Coding Freedom: The Ethics and Aesthetics of Hacking. Princeton, Princeton University Press., forthcoming.

251 See http://www.ethisphere.com/wme/; accessed on June 20th, 2012.

252 See http://www.starbucks.com/about-us/company-information; accessed on June 20th, 2012.

253 I thank Media, Peace and Conflict student Sandra Sharman, and her partner Lucian Segura, for having initiated our relationship with the plantation workers as a result of our Media Ethics class, taught by Dr. Daniela Ingruber.

254 This characterization of the "useless" was coined in conversations with liberation theologian Franz Hinkelammert.

255 Richmond, O. (2011). A Post-Liberal Peace. London, Routledge.; and Richmond, O. (2012). Failed Statebuilding Versus Peace Formation. Manchester, Manchester University.

256 Tuhiwai Smith, L. (2008). Decolonizing Methodologies: Research and Indigenous Peoples. New York, Zed Books.

257 Peer, B. (2008). Curfewed Night. New Delhi, Random House India.

Index

A

Abbottabad, Letters from 87, 155, 176, 197, 204
adhocracy 150–157, 161
Alcoholics Anonymous 113
al-Qaeda 87–90, 110, 153, 156, 205
AMAL 75, 82
Anonymous 120, 139, 140, 144, 145, 147, 148, 150–152, 153, 159, 166, 171, 172, 179, 203
Arab Spring 59, 60, 122, 153–156, 160
Assange, Julien 152, 176, 204

B

Bacha Bazi 36, 39, 151, 184
Bateson, Gregory 113–120, 125–130, 136, 158, 169, 170, 191, 199–201
bin Laden, Usama 88, 90, 118, 155
Bosnia-Herzegovina 17, 19–22, 187
brain hemispheres 125–130, 135, 153–155, 168

C

Capra, Fritjof 44, 141–144, 171, 188, 191, 202
care 20, 33, 48, 54, 55, 58, 63, 85, 99, 111, 116, 130–137, 141, 155, 164, 165
Cartesian paradigm 70, 155, 160
Chanology 146, 147
Chappel, Peter 21
community of practice 135, 142, 143, 147-156, 165
complex adaptive system 47, 78, 80, 81, 84–102, 106–112, 123, 129, 130, 137–140, 144, 155, 159, 164–167, 198
complexity 47, 71, 74, 84, 98, 104, 137, 140, 142, 151, 154
compulsive help 120
Congo, Democratic Republic of 27, 38, 45, 47, 50, 131, 171–176, 182, 188, 189

cybernetics 112, 117–121, 143, 158

D

decolonizing methodology 50
dialogue 57, 107, 138, 140, 158–161, 164–166, 186
Drvar 18–23, 172, 181
Dyncorps 36, 42

E

Easterly, William Russell 36, 38, 67, 111, 172, 185–187, 191, 198
Egypt 59, 61, 90, 145, 153, 154, 172–179, 190, 202, 205
El Salvador 52–57, 61, 67, 178, 189
Ethiopia 112, 132

F

Father Toño 54–58, 61–63, 69, 189
feedback loop 78–89, 98, 117, 121, 124, 140, 141
Fein, Peter 15, 144, 150, 153, 157, 172, 204, 205
FMLN 53, 57
Foucault, Michel 89, 178, 197
French Revolution 104

G

Gacaca 40
Gandhi, Mahatma 66, 168
Geovanni Morales 14
Gulabi Gang 95, 96, 100, 110, 130, 149, 151, 166, 168

H

Hezbollah 73–89, 110, 118, 122, 151, 169, 170, 175, 177, 193–198
human trafficking 17, 28–32, 36, 44, 53, 164, 169, 184

I

India 65, 68, 69, 92, 95, 99, 167, 176, 177, 191, 192, 197, 206
Internet freedom 145–150
Iran 75, 82, 83, 85, 89, 139, 148, 150, 169, 177, 196, 197, 203

Iraq 25, 29–37, 43, 44, 60, 67, 87, 89, 153, 156, 171–173, 178, 181–187, 197, 200, 205
Ismael, Edna Adam 133, 201

K

Kashmir 68–70, 167, 170, 175, 176, 192
kinesthetic of peace 130
Kuhn, Thomas 44, 59, 175, 188, 190

L

Leafcutter ants 141, 150
Lebanon 73, 75, 81–87, 169, 173, 175, 177, 193–196
liberal peace model 35, 39, 135
Livno 19

M

Mara 18, 53
Mara Salvatrucha 53, 67
Marceta, Mile 21
Menchu, Rigoberta 47
MINUSCO 50
Moyo 36, 177, 185
MS-13 53, 54, 56
Mubarak 145, 153, 172, 174, 190, 205
Musa Sadr 75

O

Obama, Barack 63, 120, 153, 171, 186
Occupy Wall Street 61, 122, 169, 200
open space 159

P

panarchy 100–103, 107, 110, 113, 121, 123, 128–131, 135, 136, 140, 151, 156–158, 165, 166, 198
Panzi 48
paradigm shift 25, 46–48, 52, 60, 61, 64, 70–72, 119–124, 157, 158

peace and conflict studies 17, 20, 25, 26, 38, 40, 41, 49, 52, 70, 77, 107, 111, 112, 120–122, 140, 157, 158, 161, 164, 166, 168, 189
peace mission 25, 27, 36, 39, 44, 49, 70
periodic attractor 80
point attractor 78, 80, 82, 83
poverty trap 105, 106

R

remember connection 104, 121, 125, 135, 136, 140, 156, 157, 161, 165, 166
revolt connection 104, 122–124, 136
revolution 44, 57, 58, 60, 63, 75, 80, 84, 104, 110, 122, 157, 172, 176, 178, 190
Richmond, Oliver P. 14, 35, 67, 71, 103, 107, 126, 129, 130, 165, 176–178, 184, 186, 187, 191, 192, 198, 200, 201, 205
rigidity trap 105, 106, 121
Rumi 39, 52, 61, 65, 153, 178, 186, 191

S

sacred 125–130, 136, 137, 146, 160, 161, 168
Samawada, Hiro Adam Diriye 131–137, 141, 166
Sampat, Pal Devi 90–109, 122, 126–130, 136, 142, 143, 168, 197
Sarajevo 18, 20, 22, 24
Scientology 146, 174, 203
Servicio Social Pasionista 54
Somalia 112, 131–134, 178, 187, 199, 201
Starbucks 162, 163
strange attractor 78–80, 84–90, 98, 103, 123, 141
structural violence 54, 59, 65, 70, 91, 92, 96, 107, 122

T

Tahrir Square 123, 156, 169, 175, 200
Telecomix 144, 145, 153, 159, 202
Tokmadjia Drago 20–24
torture 65, 69, 119, 132, 177, 204
turbulence 80, 82, 84, 88, 89, 98, 112, 122, 140–142, 153, 156, 164, 165

U

United Nations 36, 38, 45, 68, 171, 177, 183, 186
universality 46

University for Peace, UPEACE 14, 40, 41, 158, 163, 189, 191
UNMOGIP 68

V

Versailles, Treaty of Versailles 117, 170, 199, 200
visibility of the invisible 129, 137
vortex 80, 84, 85, 88, 89, 98, 102, 140–142, 156, 165, 198
VPN, virtual private network 144

W

Wati, Lila 97, 109, 120
Wenger, Etienne 143, 149, 179, 202
Wikileaks 147, 151, 153
wilāyat al-faqīh 82–86, 89
World Vision 133